BIOACTIVE
COMPOUNDS FROM
PLANTS

The Ciba Foundation is an international scientific and educational charity. It was established in 1947 by the Swiss chemical and pharmaceutical company of CIBA Limited — now CIBA-GEIGY Limited. The Foundation operates independently in London under English trust law.

The Ciba Foundation exists to promote international cooperation in biological, medical and chemical research. It organizes about eight international multidisciplinary symposia each year on topics that seem ready for discussion by a small group of research workers. The papers and discussions are published in the Ciba Foundation symposium series. The Foundation also holds many shorter meetings (not published), organized by the Foundation itself or by outside scientific organizations. The staff always welcome suggestions for future meetings.

The Foundation's house at 41 Portland Place, London W1N 4BN, provides facilities for meetings of all kinds. Its Media Resource Service supplies information to journalists on all scientific and technological topics. The library, open five days a week to any graduate in science or medicine, also provides information on scientific meetings throughout the world and answers general enquiries on biomedical and chemical subjects. Scientists from any part of the world may stay in the house during working visits to London.

Ciba Foundation Symposium 154

BIOACTIVE COMPOUNDS FROM PLANTS

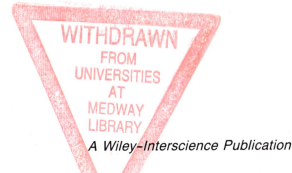

A Wiley-Interscience Publication

1990

JOHN WILEY & SONS

Chichester · New York · Brisbane · Toronto · Singapore

©Ciba Foundation 1990

Published in 1990 by John Wiley & Sons Ltd.
Baffins Lane, Chichester
West Sussex PO19 1UD, England

Other Wiley Editorial Offices

John Wiley & Sons, Inc., 605 Third Avenue,
New York, NY 10158-0012, USA

Jacaranda Wiley Ltd, G.P.O. Box 859, Brisbane,
Queensland 4001, Australia

John Wiley & Sons (Canada) Ltd, 22 Worcester Road,
Rexdale, Ontario M9W 1L1, Canada

John Wiley & Sons (SEA) Pte Ltd, 37 Jalan Pemimpin 05-04,
Block B, Union Industrial Building, Singapore 2057

Suggested series entry for library catalogues:
Ciba Foundation Symposia

Ciba Foundation Symposium 154
242 pages, 56 figures, 13 tables

Library of Congress Cataloging-in-Publication Data
Bioactive compounds from plants.
        p.   cm.—(Ciba Foundation symposium; 154)
        Editors: Derek J. Chadwick and Joan Marsh.
        Symposium on Bioactive Compounds from Plants, held in
collaboration with the Chulabhorn Research Institute at the Royal
Orchid Sheraton Hotel, Bangkok, Thailand, Feb. 20–22, 1990.
        'A Wiley–Interscience publication.'
        Includes bibliographical references.
        ISBN 0 471 92691 4
        1. Plant bioactive compounds—Congresses.   2. Plant biotechnology—
Congresses.   I. Chadwick, Derek J.   II. Marsh, Joan.
III. Chulabhorn Research Institute (Bangkok, Thailand)
IV. Symposium on Bioactive Compounds from Plants (1990: Bangkok,
Thailand)   V. Series.
QK898.B54B56   1990
660'.6—dc20   90-40978
                                                               CIP

British Library Cataloguing in Publication Data
    Bioactive compounds from plants.
    1. Drugs obtained from plants
    I. Ciba Foundation   II. Series
    615.32

ISBN 0 471 92691 4 ✓

Phototypeset by Dobbie Typesetting Service, Tavistock, Devon.
Printed and bound in Great Britain by Biddles Ltd., Guildford.

Contents

Participants

M. J. Balick Institute of Economic Botany, The New York Botanical Garden, Bronx, NY 10458-9980, USA

W. Barz Department of Plant Biochemistry, Westfälische Wilhelms-Universität, Hindenburgplatz 55, D-4400 Münster, Federal Republic of Germany

A. R. Battersby (*Chairman*) University Chemical Laboratory, Lensfield Road, Cambridge CB2 1EW, UK

D. Bellus R&D Plant Protection, Agricultural Division, R-1004.6.36, CIBA-GEIGY AG, CH-4002 Basel, Switzerland

W. S. Bowers Laboratory of Chemical Ecology, Department of Entomology, College of Agriculture, Building 36, The University of Arizona, Tucson, AZ 85721, USA

Nuntavan Bunyapraphatsara Medicinal Plant Information Center, Faculty of Pharmacy, Mahidol University, Sri Ayuthaya Road, Bangkok 10400, Thailand

R. C. Cambie Department of Chemistry, University of Auckland, Private Bag, Auckland, New Zealand

Kan Chantrapromma Department of Chemistry, Prince of Songkla University, Hat-yai, Songkla 90112, Thailand

P. A. Cox Department of Botany & Range Science, Brigham Young University, Provo, UT 84602, USA

G. M. Cragg Natural Products Branch, National Cancer Institute, Frederick Cancer Research Facility, Building 1052, PO Box B, Frederick, MD 21701, USA

E. Elisabetsky Laboratorio de Etnofarmacologia, Departamento Fisiologia-CCB, Universidade Federal do Para, 66.000 Belem Para, Brazil

N. R. Farnsworth   Program for Collaborative Research in the Pharmaceutical Sciences, College of Pharmacy, University of Illinois at Chicago, Room 309, 833 South Wood Street, Chicago, IL 60612, USA

G. W. J. Fleet   The Dyson Perrins Laboratory, University of Oxford, South Parks Road, Oxford OX1 3QY, UK

M. W. Fowler   Department of Molecular Biology & Biotechnology, Wolfson Institute of Biotechnology, University of Sheffield, Sheffield S10 2TN, UK

T. C. Hall   Department of Biology, Texas A&M University, College Station, TX 77843-3258, USA

J. B. Harborne   Department of Botany, Plant Sciences Laboratories, University of Reading, Whiteknights, PO Box 221, Reading RG6 2AS, UK

J. Kuć   Department of Plant Pathology, S-305 Ag Science Building-North, University of Kentucky, Lexington, KY 40546-0091, USA

S. V. Ley   Department of Chemistry, Imperial College of Science & Technology, London SW7 2AY, UK

Y.-L. Liu   Institute of Medical Plant Development, Chinese Academy of Medical Sciences, Haidian District, Dong Beiwang, Beijing 100094, People's Republic of China

J. MacMillan   Department of Organic Chemistry, University of Bristol, Cantock's Close, Bristol BS8 1TS, UK

Her Royal Highness Princess Chulabhorn Mahidol   Chulabhorn Research Institute, Chitralada Palace, Bangkok 10300, Thailand

L. N. Mander   Research School of Chemistry, Australian National University, GPO Box 4, Canberra, ACT 2601, Australia

E. McDonald   Chemistry Department 1, Mereside, Alderley Park, Macclesfield, Cheshire, SK10 4TG, UK

I. Potrykus   Institut für Pflanzenwissenschaften, ETH-Zentrum, Universitätstrasse 2, CH-8092 Zürich, Switzerland

R. W. Rickards   Research School of Chemistry, Australian National University, GPO Box 4, Canberra, ACT 2601, Australia

Somsak Ruchirawat   Department of Chemistry, Faculty of Science, Mahidol University, Rama 6 Road, Bangkok 10400, Thailand

W. Steglich   Institut für Organische Chemie & Biochemie, Universität Bonn, Gerhard Domagk Strasse, D-5300 Bonn 1, Federal Republic of Germany

Khanit Suwanborirux (*Ciba Foundation Bursar*)   Department of Pharmacognosy, Faculty of Pharmaceutical Sciences, Chulalongkorn University, Bangkok 10330, Thailand

M. van Montagu   Laboratorium Genetika, Rijksuniversiteit Gent, Ledeganckstraat 35, B-9000 Gent, Belgium

Y. Yamada   Department of Agricultural Chemistry, Faculty of Agriculture, Kyoto University, Yoshida, Sakyo-ku, Kyoto 606, Japan

Preface

In October 1988 the International Organization for Chemical Sciences in Development (IOCD) held a meeting in London at the Ciba Foundation. Among the distinguished participants present was HRH Princess Chulabhorn Mahidol of Thailand, who agreed to spend time after the IOCD meeting in discussions with the Foundation's staff. From this chance encounter evolved the idea of our organizing a symposium and open meeting in Thailand as a collaborative venture between the Ciba Foundation and the Chulabhorn Research Institute. It gives me very great pleasure to record publicly the help and support unstintingly given by the Princess, the Royal Consort Squadron Leader Veerayuth Didyasarin, and their staff to the Foundation during the organization and execution of the meetings.

The suggestion by Alan Battersby, Chairman of the Foundation's Executive Council, of Bioactive Compounds from Plants as a possible topic for the symposium seemed particularly apposite. Thailand enjoys a wealth of well-characterized plant life and has high standards of scientific expertise in disciplines allied to its study and use. And so it was that there assembled in Bangkok in February 1990 thirty participants from eleven countries (including five from Thailand) whose interests ranged over agriculture, biochemistry, biotechnology, botany, ethnobotany, ethnopharmacology, entomology, molecular biology, organic chemistry, pharmacology and plant pathology for the Ciba Foundation's 285th symposium (number 154 in the New Series of publications).

I suspect that all of us present during the meeting will have taken away not only cherished memories of the most hospitable of peoples, but also a deep and lasting impression of the passionate commitment of the ethnobotanists and ethnopharmacologists to the indigenous cultures with which they are so intimately involved, of opportunities tragically being wasted as the destruction of the tropical rain forests continues apace, of possibilities unrealized through an apparent lack of commercial interest in the investigation of plant-derived natural products, and of the importance of interdisciplinary communication between scientists to make best use of irreplaceable natural resources. I hope that the wide-ranging papers and discussions presented in this book will contribute in some measure to increasing awareness of these issues and of the possibilities for sympathetic utilization of the estimated 250 000 species of plants on Earth.

Derek J. Chadwick
Director, The Ciba Foundation

Introduction

Alan R. Battersby

University Chemical Laboratory, Lensfield Road, Cambridge CB2 1EW, UK

Your Royal Highness, fellow scientists and colleagues, I would like to welcome you very warmly to this symposium on Bioactive Compounds from Plants which has been arranged jointly by the Ciba Foundation of London and the Chulabhorn Research Institute here in Bangkok. It is a special pleasure to me as Chairman of the Executive Council of the Ciba Foundation that it has been possible to collaborate with our friends and colleagues in Thailand in this way. We at the Foundation are immensely grateful for all the help and support we have received from Her Royal Highness and from all the scientific and related personnel who have worked so hard with us.

The symposium is a broad ranging one which is both timely and appropriate. I say appropriate in relation to our holding the meeting in Thailand because, as many of you know, there is a very strong interest here in the chemistry of plant material and especially in bioactive substances extracted from them. History tells us what a rich collection of valuable substances has been isolated from plants. We can think of morphine from *Papaver somniferum*, emetine from Ipecacuanha, vincristine from *Catharanthus (Vinca) roseus*, colchicine from *Colchicum autumnale* and then a whole range of steroidal materials of such enormous value to the pharmaceutical industry. Probably I have not even mentioned the plant bioactive substance dear to your heart. The list is huge.

It is this history of many valuable discoveries in the past which inspires real confidence that the chances of finding new active materials are good. Vast numbers of species have still to be properly examined and modern techniques applied to plants studied a long time ago could yield pleasant surprises. After all, vincristine is a very minor component of the plant in which it occurs. So there is much work to do and this symposium will, I am sure, help in the choice of priorities regarding species and the best ways to move forward.

It is an important feature of the Ciba Foundation Symposia that they bring together scientists from many different disciplines and this one is no exception. We will be learning from botanists, pharmacologists, biochemists, structural and synthetic organic chemists, molecular biologists and experts on tissue culture. It should be an exciting meeting and my aim as Chairman will be to keep proceedings running reasonably to time and to stimulate the maximum discussion. That is where everyone has a part to play and I am sure that we will all benefit enormously from such interaction.

The role of ethnopharmacology in drug development

Norman R. Farnsworth

Program for Collaborative Research in the Pharmaceutical Sciences, College of Pharmacy, University of Illinois at Chicago, 833 South Wood Street, Chicago, Illinois 60612, USA

Abstract. There are 119 drugs of known structure that are still extracted from higher plants and used globally in allopathic medicine. About 74% of these were discovered by chemists who were attempting to identify the chemical substances in the plants that were responsible for their medical uses by humans. These 119 plant-derived drugs are produced commercially from less than 90 species of higher plants. Since there are at least 250 000 species of higher plants on earth, it is logical to presume that many more useful drugs will be found in the plant kingdom if the search for these entities is carried out in a logical and systematic manner.

The first and most important stage in a drug development programme using plants as the starting material should be the collection and analysis of information on the use(s) of the plant(s) by various indigenous cultures. Ethnobotany, ethnomedicine, folk medicine and traditional medicine can provide information that is useful as a 'pre-screen' to select plants for experimental pharmacological studies. Examples are given to illustrate how data from ethnomedicine can be analysed with the aim of selecting a reasonable number of plants to be tested in bioassay systems that are believed to predict the action of these drugs in humans. The ultimate goal of ethnopharmacology should be to identify drugs to alleviate human illness via a thorough analysis of plants alleged to be useful in human cultures throughout the world. Problems and prospects involved in attaining this goal are discussed.

1990 Bioactive compounds from plants. Wiley, Chichester (Ciba Foundation Symposium 154) p 2–21

Drugs are derived from a number of sources and in a number of ways. The earliest drugs were plant extracts, followed by natural compounds of known chemical structure, followed by inorganic substances. Since the advent of synthetic organic chemistry in the 1940s and 1950s, synthesis of compounds has been the most popular means of drug discovery. This now consumes the major portion of the research and development budgets of pharmaceutical companies worldwide. Productivity in drug development is directly proportional to resources invested. The last useful drugs to be discovered from higher plants were vinblastine and vincristine (marketed in 1961 and 1963, respectively), and the research and development budget for efforts to find new drugs in higher plants has been miniscule for the past 25 years. It is therefore not surprising that additional plant-derived drugs have not been discovered in recent times. The National Cancer Institute (NCI), over a period of about 30 years, screened more than 32 000 higher plant species for their ability to inhibit various tumours. About seven per cent of these species possessed activity and several hundred cytotoxic (*in vitro*) and/or antitumour (*in vivo*) compounds were discovered in the NCI programme. However, of about 20 of these agents that were evaluated in humans, none was effective or safe enough to result in marketed drugs. This lack of productivity in drug discovery based on a random selection of 32 000 species of plants for testing is often used as an argument against searching for other types of drugs in the plant kingdom.

The question that I shall attempt to answer in this paper is 'Can higher plants yield new therapeutic agents?' First, one must ask whether or not useful drugs have historically been discovered from higher plants.

Global importance of drugs from higher plants

There can be little doubt about the current importance of higher plants as drugs *per se* and as sources of useful drugs on a global basis. It has been estimated that 80% of people living in developing countries are almost completely dependent on traditional medical practices for their primary health care needs, and higher plants are known to be the main source of drug therapy in traditional medicine. Since about 80% of the world's population resides in developing countries, about 64% of the total population of the world utilizes plants as drugs, i.e. 3.2 billion people (Farnsworth et al 1985). If we are interested in drug development using higher plants as potential leads, it is obvious that the combined experience of the 3.2 billion users of plants as drugs must be considered. On the other hand, there are sceptics who believe that this human experience has already resulted in the discovery of as many 'Western' style drugs as will ever be found.

On the basis of global survey data we have found that about 119 plant-derived chemical compounds of known structure are currently used as drugs or as biodynamic agents that affect human health. Less than a dozen of these are produced commercially by synthesis or by simple chemical modification of the extracted active principles; the remainder are extracted and purified directly from plants. These 119 useful drugs are obtained from only about 90 species of plants (Farnsworth et al 1985).

In the United States, the only developed country where data are available, we found that from 1959–1980, 25% of all prescriptions dispensed from community pharmacies contained active principles that were extracted from higher plants. This figure did not vary by more than one per cent in any of the 22 years surveyed. In 1980, consumers in the United States paid more than

$8 billion (US) for prescriptions containing active principles obtained from higher plants (Farnsworth & Morris 1976, Farnsworth & Soejarto 1985, Farnsworth et al 1985).

Many of the plant-derived drugs currently in use are prototypes. Any textbook of pharmacology has morphine as the major topic in any chapter on analgesics or atropine in a chapter on anticholinergic agents. The most important of the plant-derived drugs are (alphabetically) atropine, bromelain, caffeine, chymopapain, codeine, colchicine, deslanoside, digitoxin, digoxin, emetine, ephedrine, lanatosides A, B and C, morphine, papaverine, physostigmine, pilocarpine, pseudoephedrine, norpseudoephedrine, quinidine, quinine, reserpine, scopolamine, sennosides A and B, theophylline, tubocurarine, vinblastine and vincristine. On the basis of current usage and the inestimable role that these drugs have played in the alleviation of human suffering, it seems justified that the search for new and improved plant-derived drugs should continue. It should also be pointed out that many biologically active plant-derived substances that have not been useful as drugs *per se*, have been models for synthesis or structural modification to produce useful drugs. For example, khellin was used for many years to treat angina and as a bronchodilator but had many side effects, including nausea and vomiting. Using this compound as a model, the more useful and less toxic cromolyn was synthesized (Buckle 1984). Mescaline, the cannabinoids, yohimbine and many other plant-derived biodynamic agents have been important experimental tools in pharmacology.

Approaches to drug discovery using plants as starting material

It is beyond the scope of this paper to identify all of the approaches that have been or might be used to identify plants with the greatest potential as useful drugs. There are two basic problems that have to be addressed in any drug development programme using plants as starting material. The first is the ability to acquire well documented plant material that has the greatest possibility of possessing the biological activity being sought. The second is the ability to assay extracts from the plants acquired. It is my personal opinion that industry has, until the past three or four years, neglected plants as a source of potential drugs because they have been unable (or unwilling) to acquire plant materials from those areas of the world where the greatest biodiversity exists or where information is available indicating that the plants contain bioactive agents. Reasons for this neglect no longer exist: distance is not a problem, the world is a very small place; information that can be used to predict which plants may contain bioactive agents is available; collecting and analysing this information is discussed below.

Until recent times, most bioassays were carried out on live animals. In developed countries there are political pressures to discourage the use of animals for research, especially in the early stages of drug development. A disadvantage of *in vivo* models is that there is a limit to the number of parameters that can be measured in a single experiment. Setting up animal models to detect properties such as antidiabetic, anticancer, antiinflammatory, antiageing, antihypertensive and immunomodulatory activities is time consuming and expensive. Indeed, for some biological activities there may not be a good correlation between results from animals and those from humans. Animal models are also expensive, they are subject to considerable variation, less quantitative than one would desire, not usually mechanistically driven and results vary considerably depending on the route of administration of the substance being tested. In the area of drug development from plants, once a crude extract is discovered that has an interesting biological activity, the bioassay model must be used to monitor a chemical fractionation until the active principle(s) is (are) discovered. This becomes even more expensive and requires large amounts of plant extract, which may not be available in the early stages of the discovery process.

Interest has been rekindled within certain segments of the pharmaceutical industry in recent years owing to the innovative development of *in vitro* bioassays based on inhibition or stimulation of receptor binding, inhibition or stimulation of biochemical reactions and/or inhibition or stimulation of enzyme systems. These types of assays are mechanistically driven, require very small amounts of plant material, are usually simple to perform and the procedures are amenable to automation, thereby reducing costs. It is not uncommon to see an industrial laboratory with a capability to conduct *in vitro* bioassays on 1000–5000 different samples per day.

From an industrial point of view, the major problem today is the acquisition of large numbers of plants at a reasonable cost. Conservative estimates are that about 250 000 species of higher plants live on this planet. It would be impossible to collect representative samples of all of these species for any drug development programme. This leads us to consider the need for selective collection of plants that have a prior history of use by humans for specific pathological conditions. In one sense, this process can be considered as a 'pre-screen': the hypothesis is that if a plant is documented as having been used by humans for centuries, some degree of credibility must be given to its efficacy, or the plant would not have been continuously used for such long periods. This brings us to a discussion of a complex set of terms representing areas where useful information on the medicinal use of plants might prove helpful.

Ethnopharmacology and the potential value of ethnomedical information

According to Rivier & Bruhn (1979), ethnopharmacology is a multidisciplinary area of research, concerned with the observation, description and experimental investigation of indigenous drugs and their biological activities. Bruhn & Holmstedt (1981) later modified the definition to 'the interdisciplinary scientific exploration of biologically active agents traditionally employed or observed by man'. Ethnopharmacology, as a scientific term, seems to have been introduced at an international symposium held in San Francisco in 1967 to highlight the historical, cultural, anthropological, botanical, chemical and pharmacological aspects of traditional psychoactive drugs (Efron et al 1967).

The 'observation' part of this description is most often carried out by ethnobotanists in the field; less often (based on reports in the scientific literature) by medical doctors, biologists, pharmacognosists, pharmacologists, chemists and a variety of other interested scientists, or by lay people. An ethnobotanist is interested not only in the uses of plants by indigenous cultures (ethnomedicine) but also in any other economic uses of plants. Folklore medicine seems to relate to observations or claims by (usually) scientifically untrained persons that a material or ritual produces a medically useful effect. Where does the term 'traditional medicine' fit within the confusing terminology of ethnobotany, ethnomedicine, ethnopharmacognosy, folk medicine and ethnopharmacology? According to the World Health Organization (WHO), 'traditional medicine' is a vague term loosely used to distinguish ancient and culture-bound health care practices which existed before the application of science to health matters in official modern scientific medicine or allopathy. Some frequently used synonyms, according to the WHO, are indigenous, unorthodox, alternative, folk, ethno-, fringe and unofficial medicine (Bannerman et al 1983). The major 'traditional' medical systems seem to be Chinese traditional medicine, Ayurveda and Unani, at least based on the number of recipients of health care under these systems, which are prevalent mainly in China and in India, but are also used in other countries. I personally consider that these traditional medicinal systems embrace the following elements of 'credibility': (a) they have a long historical written documentation, (b) they are based on theories, (c) they are based on formal educational systems, and (d) there is evidence of periodic revision of the systems based on research. Plants used in these types of formal medical systems should have a greater chance of providing useful 'Western' style drugs than plants collected at random and then tested. Information on the medicinal uses of plants collected by ethnobotanists and others directly from practitioners or users of non-formal systems of medicine, if properly recorded and documented, would also be more productive with regard to drug development than would be the random collection screening process.

Our global survey has shown that about 74% of the 119 useful drugs extracted from higher plants that are in current use were discovered by chemists who were studying plants to elucidate the principle(s) responsible for the ethnomedical claims attributed to the plants (Farnsworth et al 1985). This obviously is not meant to imply that 74% of ethnomedical data will lead to useful drugs.

Sources of ethnomedical information

The ethnomedical literature is not abstracted in a systematic manner and journals and books presenting information on the alleged medicinal applications of plants are difficult to locate in a simple way, even though many exist. Almost every country of the world is covered by one or more books, primarily written by botanists, with titles such as 'Medicinal Plants of the Philippines' or 'Medical Botany of Nepal'. Papers representing compilations of information on the ethnomedical uses of plants are scattered throughout a number of journals, primarily those published in developing countries, but some in recent years in the *Journal of Ethnopharmacology*. I find particularly useful citations to ethnomedical uses of plants in 'Medicinal and Aromatic Plants Abstracts', published by The Publications & Information Directorate, CSIR, Hillside Road, New Delhi 110 012, India.

Chinese literature on medicinal plants is computerized at the Chinese Medicinal Material Research Centre, The Chinese University of Hong Kong and the database is accessible off-line. Starting in 1987 this group has published Abstracts of Chinese Medicines, each issue consisting of about 300 pages of abstracts of work on Chinese medicinal plants.

At the University of Illinois at Chicago, we initiated an ambitious programme to computerize the world literature on natural products, starting in 1975. Our system is referred to as NAPRALERT (Natural Products Alert) and is available both on- and off-line (Farnsworth et al 1981, Loub et al 1985). We computerize the ethnomedical, pharmacological and chemical aspects of all natural products: approximately 85 000 articles had been computerized by February 1990. Since the system represents a fully relational database, it is possible to analyse data on the medicinal potential of plants. At the end of 1988 we had information on about 26 000 species of higher plants; about 41 940 records of ethnomedical uses of plants are in the database. Table 1 presents the most frequently documented uses claimed for plants in the database.

If one excludes data from Table 1 relating to reproduction, i.e. contraceptives, emmenagogues, abortion-inducing agents, and ecbolic compounds or those that promote fertility, in which we have had a special interest on behalf of the World Health Organization and for which we thus have made an exhaustive search, the major uses of plants by indigenous cultures relate to health problems that are prevalent in developing countries. These include infectious diseases, parasitic diseases, diarrhoea (often related to parasitic diseases), fever, impotency, jaundice (for which it is difficult to distinguish the aetiology, e.g. hepatitis B, alcohol consumption or use of plants as medicines that might contain hepatotoxic agents), and the common cold.

Problems related to interpretation of ethnomedical data

Botanists are the main providers of ethnomedical information. However, most botanists have no or limited training in medicine or pharmacology. For a scientist interested in using this information to select plants for scientific studies, I find many problems with the manner in which they collect data. An ethnomedical

TABLE 1   Frequency of ethnomedical records in the NAPRALERT database[a]

Medical condition for which used or other application	No. records	Medical condition for which used or other application	No. records
Menstrual induction	4110	Epilepsy	299
Induction of abortion	2630	Genitourinary problems	298
Reduce inflammation	1879	Kill fish	294
Prevent post-partum haemorrhage	1547	Insecticide	289
		Vermifuge	265
Bacterial diseases	1521	High blood pressure	264
Induction of diuresis	1327	Produce emesis	260
Reduce fevers	1299	Induce sweating	256
Impotency	1275	Snake and spider bites	250
Reduce pain	1255	Cardiotonic	224
Contraceptive	1249	Carminative (treat bowel wind)	214
Laxative/cathartic	1032		
Diarrhoea	922	Produce hallucinations	213
Anthelminthic	867	Tuberculosis	178
Antispasmodic	856	Prevent emesis	146
General tonic	749	Treponemal diseases	145
Jaundice/liver disorders	733	Haemorrhoids	145
Stimulate lactation	629	Fungal infections	142
Reduce cough reflex	621	Emollient	118
Dysmenorrhoea	620	Produce narcosis	115
Sedative	606	Stomach ulcer	110
Accelerate wound healing	576	Increase bile flow	107
Malaria	565	Schistosomiasis	101
Cancer	552	Common cold symptoms	98
Digestive aid	540	Taenifuge (tapeworm)	91
Female 'problems'	516	Trypanosomiasis	87
Diabetes	514	Hair stimulant	84
Expectorant	473	Bitter (appetite stimulant)	80
Fertility promotion	432	Dissolve kidney stones	79
Food use	414	Anticoagulant	78
Asthma	410	Haematinic	71
Astringent	380	Burns	70
Haemostatic	323	Tranquilizer	68
Stimulation of central nervous system	321	Escharotic (caustic)	67
		Prevent miscarriage	63
Viral diseases	315	Repel insects	62

[a]Based on a total of approximately 300 types of ethnomedical uses involving 41 940 computer records.

report frequently provides only a single word for the use of a plant; for example, the plant is used as a 'contraceptive'. They do not seek further information of importance, such as (a) Is the plant used by the male or the female? (b) If used by the female, is it used on a regular cycle basis? (c) Is the user of child-bearing age? (d) How much is used? (e) How is the medicine prepared? (f) Does the user have more than one sexual partner (if only one, could he/she be infertile?)? (g) Has the woman had children (she may be infertile)? Answers to these questions are important in the decision about whether to risk setting up an assay that might cost over $1000. More importantly, such information would enable a proper experimental model to be established.

Of more than 1275 records we find in NAPRALERT suggesting that various plants are used as aphrodisiacs, i.e. for impotency, all but about 25 fail to indicate whether this effect occurs in the male, the female or both. Along similar lines, one often finds literature indicating that a plant is used as a 'stimulant'. Without further documentation, one is at a loss to attribute this effect to stimulation of the central nervous system, as with *Catha edulis* (a plant containing amphetamine-like constituents), or to a simple feeling of warmth or digestant action on foods.

A perusal of any recent issue of the *Journal of Ethnopharmacology* will illustrate the problems that face one interested in drug development. Typically, a paper reporting the ethnomedical uses of many plants from a specific geographic area will provide the reader with a table of data giving the Latin binomial of each plant, usually the common name(s), the plant part(s) used (often not included), and single words describing the use(s). Less frequently one will find the type of preparation used, e.g. infusion, decoction, paste.

Another major problem in collecting ethnomedical information of optimal value for drug discovery purposes is that of language and translation. I can personally recall, during a visit in the People's Republic of China in 1974 as a member of the Herbal Pharmacology Delegation from the United States, asking a traditional practitioner about the use of a plant that was growing in a medicinal plant garden. He indicated to me that the plant was used 'for fevers'. Later that evening in a discussion session of our delegation, it was discovered by our Chinese language expert that the practitioner meant that the plant was used 'to remove heat from the body', which in Western thought would be to reduce fever. In reality, the plant was used for liver disorders. Cox (1990a, this volume) has been studying the ethnopharmacology of Samoans and has eloquently described the problems inherent in real ethnomedical research, in which a mastery of the language is imperative to equate the medical terminology of the disease states with that of our Western concept of medicine.

It is strongly recommended that before any programme is initiated to look for new bioactive compounds in plants on the basis of an analysis of ethnomedical data, the excellent papers of Croom (1983) and Lipp (1989) should be consulted.

As one interested in evaluating ethnomedical data to obtain a short-list of plants for specific biological testing, I am often frustrated by the lack of essential details on what conditions are being treated and the methods of use of ethnomedical preparations in reports from the people collecting ethnomedical information. I would suggest that most ethnobotanists should concentrate more on details in their collection of information on plant uses, if they expect their work to be beneficial and to lead ultimately to drugs for the benefit of humankind.

Summary and conclusions

There is little doubt that plants have provided, and can continue to provide, humans with useful drugs. While a mass random collection of plants, followed by subjecting them to a 'shotgun' mixture of bioassays, would eventually provide leads that would result in new drugs, this approach is irrational and expensive. It seems more logical to benefit from human experience in the past and present by selecting plants that are claimed to be useful for the alleviation of human illnesses, then to subject extracts of these plants to specific, mechanistically based bioassays to substantiate their effects in people. To do this, we need more true ethnobotanists or medically trained persons with botanical expertise, to collect 'new' information from current users of plants before this information disappears forever. Mark Plotkin has said that the death of a 'medicine man' in remote areas of South America is not unlike the loss of a library, i.e. his information dies with him. The information collected and disseminated by the ethnobotanist, or analogous scientist, must also be analysed in a systematic manner to eliminate from the selection process plants used by primitive people that are known to contain chemical compounds that can explain the medical value of the plant, but which, for one reason or another, would be unsuitable for drug development.

Only a fool would neglect to investigate further the abundant flora of the world for the presence of new drugs that will benefit not only the world, but also the fool.

References

Bannerman RH, Burton J, Ch'en WC (eds) 1983 Traditional medicine and health care coverage. World Health Organization, Geneva
Bruhn J, Holmstedt B 1981 Ethnopharmacology, objectives, principles and perspectives. In: Beal JL, Reinhard E (eds) Natural products as medicinal agents. Hippokrates Verlag, p 405–430
Buckle DR 1984 Disodium cromoglycate and compounds with similar activities. In: Buckle DR, Smith H (eds) Development of anti-asthma drugs. Butterworths, London, p 261–296
Cox PA 1990a Samoan ethnopharmacology. In: Wagner H, Farnsworth NR (eds) Economic and medicinal plant research. Academic Press, London, vol 4:122–139

Cox PA 1990b Ethnopharmacology and the search for new drugs. In: Bioactive compounds from plants. Wiley, Chichester (Ciba Found Symp 154) p 40–55

Croom EM Jr 1983 Documenting and evaluating herbal remedies. Econ Bot 37:13–27

Efron DH, Holmstedt B, Kline NS (eds) 1967 Ethnopharmacological search for psychoactive drugs. Publ 1645, US Department of Health, Education and Welfare. Government Printing Office, Washington DC

Farnsworth NR, Morris RW 1976 Higher plants—the sleeping giant of drug development. Am J Pharm 148:46–52

Farnsworth NR, Soejarto DD 1985 Potential consequence of plant extinction in the United States on the current and future availability of prescription drugs. Econ Bot 39:231–240

Farnsworth NR, Loub WD, Soejarto DD, Cordell GA, Quinn ML, Mulholland K 1981 Computer services for research on plants for fertility regulation. Korean J Pharmacogn 12:98–109

Farnsworth NR, Akerele O, Bingel AS, Soejarto DD, Guo ZG 1985 Medicinal plants in therapy. Bull WHO 63:965–981

Lipp FJ 1989 Methods for ethnopharmacological field work. J Ethnopharmacol 25:139–150

Loub WD, Farnsworth NR, Soejarto DD, Quinn ML 1985 NAPRALERT: computer handling of natural product research data. J Chem Inf Comput Sci 25:99–103

Rivier L, Bruhn JG 1979 Editorial. J Ethnopharmacol 1:1

DISCUSSION

Battersby: You referred to the NCI anticancer screen; looking at those numbers it's rather a depressing picture—large numbers of plants have been examined with no useful products at the end. But the target is an incredibly difficult one and one could say that this outcome arises simply from the difficulty of the target.

Farnsworth: At the time of the old NCI programme, there was less known about the disease cancer. As the programme developed it became obvious that cancer is 200 different diseases or more. You can't expect that by using one mouse leukaemia in the screen you will find a magic bullet that will kill all cancers. Unfortunately, that took many years to find out. In the new NCI programme, they believe they have overcome this with a new protocol for screening both natural products and chemically synthesized compounds. This involves testing everything for *selective* toxicity against about 65 human tumours in cell culture.

Cragg: I would like to take this opportunity to make a few comments about the NCI programme. This is, of course, aimed at the discovery of new agents for the treatment of cancer and AIDS. The focus is on natural products; we are looking for new leads.

There are three collection contracts for plants, including one with the New York Botanical Garden, for which Michael Balick is the principal investigator. There are an extraction laboratory and a repository, where the extracts are stored at −20 °C, as well as screening laboratories for cancer and AIDS, and chemistry labs trying to isolate the active agents.

The plant collections are focused in South East Asia—including Thailand, where we have very good collaborations with local botanists—Central Africa and Madagascar, and Central and South America. There are also marine organism collections, mainly from the Great Barrier Reef area off Australia, off the coast of New Zealand, and round Papua New Guinea. There are also various microbial projects.

In the plant collections, the focus is on taxonomic diversity, including medicinal plants, but in terms of traditional medicine cancer is a poorly defined disease so we can't focus purely on the ethnobotanical leads. Separate plant parts are collected; generally the bark and the roots are the major sources of bioactive compounds. The programme includes collaborations in medicinal plant investigations—Kunming Botanical Research Institute in China, the National Products Research Institute, directed by Professor Woo, at Seoul National University, and Paul Cox at Brigham Young University, Utah. An interesting new structure type with anti-HIV activity has been isolated from a plant extract supplied by Paul Cox; this is still under patent consideration. It is not a novel chemical structure, but it is novel from the point of view of anti-HIV activity.

NCI is hoping to form collaborations with the Chinese Academy of Medical Sciences, the Brazilian Foundation for Medicinal Plants—Elaine Elisabetsky being a key person there. I have just visited India, where we are trying to establish some collaborations.

The strategy for investigation of natural materials is to obtain organic and aqueous extracts, as Michael Balick will describe (p 22). The aqueous extraction is important because in many traditional medicine systems the preparation used for treatment is usually aqueous.

The crude extracts are screened, and the active extracts are fractionated using bioassay-guided separation procedures in either the cancer screen or the AIDS screen. Once a pure active compound has been isolated further development is undertaken. All extracts are stored at $-20\,^{\circ}\text{C}$. At the moment there are 12 double storey walk-in freezers; eventually we estimate we will have of the order of a million samples derived from up to 100 000 natural materials stored. These are being kept for future screens too. The NCI goal is to find new drugs for the treatment of cancer and AIDS. Many groups would like samples of extracts, but we intend to store these for new AIDS and cancer screens as they are developed. Unfortunately, the size of the samples does not permit their ready distribution for other bioactivity screens, unless they are related to the main NCI mission.

The anticancer screening strategy reflects the complex nature of cancer, which comprises up to 200 different disease types. Even within a single disease type there's a great heterogeneity of cells. NCI previously used one single cancer, the P388 mouse leukaemia, as the primary screen. Of course, a number of agents which had good antileukaemic activity were discovered, but very few agents active against the refractory solid tumours of colon, lung, breast, ovary, prostate and CNS, or melanomas. The new screen comprises an *in vitro* primary screen employing human tumour cell lines representative of the major solid tumour disease types. Any material found to be active in the *in vitro* screen is then tested *in vivo*, using xenograft models. If the activity is confirmed, we proceed to formulation toxicology and further development.

The panels of cell lines include leukaemias, various lung lines, renal, CNS, colon, ovarian and melanoma lines. At the moment we have 60 different cell lines, against which every extract and every compound is screened, generally in triplicate at 4–5 different concentrations. The aim is to screen 10–20 000 materials a year, amounting to over 12 million assays a year. Much of the screening is automated and uses a spectrophotometric end-point to measure cell-kill.

The anti-HIV screen will be described by Michael Balick (p 27); close to 30 000 materials a year are currently being tested in this screen.

The NCI's attitude to compensating countries where raw materials, such as plants and marine organisms, are collected is as follows. In the short-term, NCI will provide the results of screening to the collection contractors, New York Botanical Garden etc, and they then distribute those results to the countries where the plants were collected. NCI requests that confidentiality be maintained because, obviously, these results are very preliminary. We wouldn't want it advertised that a certain plant had shown preliminary anti-HIV activity, resulting in an invasion of the country concerned to collect that plant material. And there are patent considerations. Local scientists and officials are invited to visit NCI to see the facilities. This is very important because these officials and scientists have a say in whether collection permits will be granted. We like to discuss the programme with them and show them that NCI intentions generally are to collaborate and share results with them. Local scientists are invited to work in NCI or NCI contract laboratories, particularly in the chemical isolation of active agents. We try to provide assistance to local scientists in upgrading facilities, mainly herbaria. Many of the herbaria in the developing countries are not adequately equipped for storage. The provision of such assistance by NCI to a foreign institution is not easy, but can be justified in terms of benefit to the NCI programme and the local institution.

Last, but not least, if a drug is isolated and patented by the US Government, in licensing that drug to a pharmaceutical company NCI requires that a certain percentage of the royalties from the sales of that drug be returned to the country of origin of the source material. The percentage will vary from drug to drug. It is not likely to be a large percentage, and, if a drug were developed from a natural product lead, the percentage would be somewhat lower. If royalties from sales of a drug amounted to $100 million a year, however, that could mean one or two million dollars to the source country, which would be a real boost to their drug development programme.

We do stress that development of a drug is a 10–20 year process, so instant returns cannot be expected. To find a good anticancer drug, the chances are, being optimistic, one in ten thousand. But I think it's important to hold out the prospect that, should a good drug be developed from one of these plants, the source country will obtain some benefit.

Bellus: Professor Farnsworth, I was very impressed by the numbers you presented. You said that only about 17% of all new natural product structures described in the literature are tested in bioassays. I think that figure should be much higher but laboratories usually do not report their negative results.

There is another hurdle, namely that very often these products are not available in quantities sufficient for bioassays. Therefore I would like to make a plea to all natural product chemists—don't just isolate small amounts which are enough for NMR or X-ray analysis, but tangible amounts sufficient for biological investigation.

Battersby: Could you put a figure on this tangible amount?

Bellus: In our field (plant protection agents) we can do about 30 microtests with 200 mg of a given compound. In the last 15 years we have tested about 2500 natural products and discovered one useful lead compound, a derivative of which is on the market already.

Hall: In your figures of the extracts from plants used in traditional medicine that were screened for antitumour activity, 136 species of plants gave an 18% hit rate. Can you tell us more about any of those extracts, for example their activity or the type of compound identified?

Farnsworth: The picture is not as clear cut as these numbers suggest. For example, at the time that most of these screens were done, any plant containing a cardiac glycoside gave a positive test, most plants with saponins gave a positive test, some sterols gave a positive test, depending on the tumour system or cell line that was being used. None of those classes of compounds would be useful clinically. Saponins kill fish, so you get a high hit rate from plants that have a reputation as a source of fish poisons but you already know that probably the compound is not going to be useful as a drug.

I think that folkloric reports of antihelminthic activity are more useful. Even if a native is difficult to converse with and doesn't know Western medicine, he does know that when he has a bowel movement he can see 'worms' in the faeces. If he then takes a plant extract and notices that the 'worms' disappear his 'knowledge' of the use of such plants is useful. That's almost as good as a human bioassay as far as I am concerned. But when a native tells you he has pains in the stomach, what does that mean? He could have stomach cancer, peptic ulcers, gas, or many kinds of parasitic infections. If somebody collects information and says that a particular plant is used for pains in the stomach, it doesn't tell me much in terms of predicting biological activity or the type of bioassay required to test for that activity. As I mentioned for compounds used as aphrodisiacs or contraceptives, we need detailed information. It takes a long time to gather that kind of information and very often the people are not willing to give it. But when you do get that kind of information, it is valuable as a predictive tool.

Balick: I was also impressed by the numbers—you said 39% of fish poisons showed antitumour activity, 52% of arrow poisons. I do a lot of work with poisonous plants—those used to prepare fish poisons, animal poisons and poisons used for assassination. Given that the nature of the screen has changed, do you think toxicity is a viable strategy for targeting activity in the *in vitro* screens of the NCI?

Farnsworth: I think those types of things will, across the board, be cytotoxic but you won't see selective cytotoxicity and therefore those extracts would not be currently of interest to NCI for further development. I would exclude these leads if I knew about them during the collection process.

Kuć: I am intrigued by the search for natural products among tribal folk medicines. I can clearly see a role in preventive medicine for antimalarial drugs or contraceptives. However, I also see tremendous complexity in bioassays for preventive drugs based on information derived from tribal medicine men. How do you assay for preventing a pain in the stomach? How do you assay for a preventive for at least 200 different types of cancer?

Farnsworth: Usually, when I give a talk like this, the last slide I show is a Las Vegas crap table, where you shoot dice and you gamble. All drug development, no matter how well is has been theoretically conceived, is a big gamble. Probably more drugs have been developed through serendipity than through any planned attack. But serendipity is educated observation of abnormal events that are picked up and followed through. The discovery of vinblastine and vincristine was serendipitous.

There were two groups working on the plant *Catharanthus roseus* simultaneously—Noble, Beer and Cutts in Ontario and Gordon Svoboda at Eli Lilly Company in Indianapolis—unknown to each other. The plant was selected and collected by Svoboda from the Philippines and by Noble, Beer and Cutts from the West Indies.

C. roseus was used as an oral insulin substitute in the Philippines during World War II when insulin became unobtainable. In the report by Noble et al (1958) on the discovery of vincaleuroblastine (vinblastine) their introductory comment was 'The results of our research which are presented here in detail for the first time, should not be considered in terms of a new chemotherapeutic agent, but rather in terms of a substance with potential chemotherapeutic possibilities'!

They made an extract of the plant material, injected it into a few rats; in three days all the rats were dead—therefore they rationalized this was a very toxic extract. So they diluted the extract and injected a few more rats and in three days all the rats were dead, so this was obviously a very very toxic extract. So they diluted the extract again and injected some more rats and three days later, lo and behold, all the rats died. Well, either they ran out of rats, or they became curious about why the deaths occurred.

They autopsied some of these rats and found that they had died from a *Pseudomonas* infection. There was no *Pseudomonas* in the plant extract so where had it come from? This is a common organism in animal houses; it is excreted in the faeces of laboratory animals. They then took white blood cell counts from a new set of rats. After one injection of plant extract, the white blood cell count decreased dramatically and then returned to normal. They rationalized that in leukaemia there is a proliferation of white blood cells and this was a substance that destroyed white blood cells. Hence, this was seen as of potential value in leukaemia.

They went through a very complex bioassay (leukopenia-directed fractionation of the plant material) and ended up with a few milligrams of crude vinblastine.

At the same time at the Eli Lilly laboratories in the USA, Svoboda selected 240 plants for his programme. *C. roseus* was the 40th plant. He submitted the sample for testing and found that in P-1534 leukaemic mice there was a prolongation of life greater than 20 days. He isolated vinblastine probably shortly after the Noble group. He noticed that in the crude extracts, after he took out the vinblastine, there was a substance that prolonged the life of these leukaemic mice for several weeks (a few were laboratory 'cures'), and eventually isolated vincristine. Both of these drugs took less than two years from the time he acquired the plant material to the time they were approved by the FDA for use on humans in the US.

Vincristine, at the beginning, was only obtained in yields of 30 g from 15 tonnes of dried leaves. They commissioned several high-powered chemists in the US to find a way to convert vinblastine to vincristine. Vinblastine is a major alkaloid in the plant, a thousand times more abundant than vincristine, and the only difference between them is an *N*-methyl group on vinblastine and an *N*-formyl group on vincristine. The leading indole alkaloid chemists in the United States tried for three years to do this and failed. A Hungarian chemist by the name of Kalman Szasz in Budapest did this by a chromic acid oxidation at $-70\,^{\circ}\text{C}$ with a 98% yield. So Kalman Szasz's efforts brought more hard currency into Hungary at that time than any other product.

One has to plan a scientific agenda within the limits of time, money, people, equipment and so forth that will cover the broader spectrum of what you are looking for. Then you implement the plan and see what happens. In cancer you have to stop abnormal cell growth; how selectively you do that determines whether you succeed or not.

I don't think people who are looking for insect feeding deterrents run into that same problem. The assay system is less expensive and much simpler. In mammalian systems you are looking at effects on enzymes and receptor sites and you have to hope effects seen *in vitro* will also be active *in vivo*.

Kuć: My question was really directed at the preventive aspects. As I understand it, vincristine is curative. I find it difficult to comprehend a screen for compounds that inhibit cell division or cell growth as an assay for preventing cancer.

Farnsworth: One has to look at this in the broadest terms. People are dying from cancer and there has to be an effort to do something not only to alleviate the pain but hopefully to stop the cancer before the terminal stage. That's one aspect of the NCI programme.

The other aspect is chemoprevention, which has only come into being in the last five years. I could put a menu on the board of all kinds of vegetables, fruits and common foods, every one of which contains compounds that in bioassay systems produce cancer preventive effects. The NCI in the US has a programme for chemotherapeutics and one for diagnostic improvements and so forth. There is not just a search for cytotoxic materials that will kill cancer cells.

Bowers: Much of the money from the NCI is targeted to finding compounds that interfere with some aspect of carcinogenesis. Far more people die of malaria, of filariasis, even of starvation, than die of cancer. I can't contain my enthusiasm for the development of databases that would link medicinal practices of one kind or another with natural products.

However, much of the secondary chemistry of plants is focused on defence against herbivorous insects and plant pathogens. The search in plants for anticancer drugs is really a very long shot in the dark. Scientists with mechanism-directed assays focused on the detection of plant chemical defensive compounds should be included in those allowed to examine these plant resources. I believe that the discovery of novel chemistry in plant extracts, of compounds targeted to the protection of plants, have a higher chance of success and ultimately would have far more impact on the quality of life.

The database is an exciting achievement, but it is only as good as those who create it, and it will focus only on those attributes in which the creator is interested. Your database tries to cover everything, including insect repellents, insect growth regulators, fungicides and natural herbicides—that's wonderful. If, at the same time, this database could indicate individuals who could most take advantage of the relationships the database identifies, that would broaden the exploitation of these natural products.

Cox: What Professor Bowers is suggesting should culminate in a repository of extracts. Too much work is done where plants are collected with the goal of meeting a specific target. For example, the National Cancer Institute programme collected these 32 000 plants that they tested against a mouse cell leukaemia system. Many of the plants they collected could be used today in other sorts of bioassays.

Bowers: This collection is an awesome effort and represents a tremendous expenditure of funds. If through this database we can identify people with mechanism-directed assays that are not covered in any other programme, they could share in this enormous resource of natural products. Then I think we would make many important discoveries and broaden the use of resources.

Mander: What is the availability of this database and what do you need to run it?

Farnsworth: We have an agreement with the World Health Organization to make this available, free of charge, to scientists or institutions in the developing countries. They write to us, say what they want and we send it back by airmail. At present, in the US the database is on-line. Outside the US the system can be used off-line. For example, if you wanted a list of all plants or extracts which show antimalaria-like activity it would cost about $10 for the first three pages and 50 cents a page thereafter. We are about to sign a contract with Chemical Abstracts Service to put NAPRALERT on-line worldwide. There is a sign up licence fee—an individual pays $100 a year—then there are some computer charges and so forth. For graduate students who need some help, we waive the fee. For a legitimate scientist who's writing an academic book that our group feels has some value, we give them data for free or for a reduced charge. Commercial users are charged higher fees.

Cox: Professor Farnsworth started by arguing that we should look towards plants as a source of useful natural products. He made a rather pointed attack on random screening and showed that even though natural products are useful there is a very low rate of success with random screens. Professor Farnsworth proposed that the NAPRALERT database or other similar efforts might be a good way of identifying plants that have had a history of folk use. He then made a very convincing argument that plants that have a folklore use are more likely to generate useful bioactive molecules than those derived from a random screen.

The NAPRALERT database is extensive but it does not distinguish between careful ethnobotanical work and casual reports on floras generated by scientists not conversant with the language of the indigenous people. So I would like to suggest a further refinement. Scientists should go out and talk to indigenous peoples who are still practising ethnomedicine. These scientists should speak the languages of the indigenous peoples and record their knowledge, so we don't have the level of resolution as 'good for stomach pains' as reported by somebody that the investigator contacted in the market, but a very specific interview with the healer who is proficient in treating different diseases. Do you think that was a good summary and do you think there is a need for more field work in ethnopharmacology?

Farnsworth: I doubt there are more than a dozen ethnobotanists in the world whose serious intent is this type of work. There needs to be more training programmes; I am unaware of any academic institution where one can study for a PhD in ethnobotany or in ethnopharmacology. People who are working as ethnobotanists are usually people trained in botany for whom ethnobotany becomes a hobby. There needs to be more people with a better medical background trained to learn from native peoples. Your work in Samoa is classical—you ran into the problem with the language and understanding the language and translating it into terms familiar to Western medicine.

What we are all saying is that a lot of the ethnomedical information has not been collected in a proper way or has not been assessed in a proper way. This type of research is the least expensive type of research in drug development and more of it should be done, especially before the information and the plants are lost for ever.

Cambie: There is a relatively small number of new compounds that are obtained yearly and tested for activity. You suggested, Professor Farnsworth, that this might be due to the fact that people didn't have the expertise or the facilities to do this testing. Have you got any suggestions how we might improve that situation?

Farnsworth: I have a simple solution. Some international organizations, such as UNESCO and the World Health Organization, should get together and set up an appropriate institute for natural product testing—Thailand would be an ideal place—an international cooperative institute for biological assessment of natural products and synthetics as well. Many people are producing compounds and not testing them in more than one assay. This could be done at a reasonable price; you couldn't cover every possible bioassay but you could find a nucleus of assays that would provide a great many leads for drug development.

Mander: We are informed that approximately 119 useful drugs of known structure have been isolated from plants. I wonder whether Professor Farnsworth can provide us with an estimate of the number of synthetic or semi-synthetic drugs which have been inspired by the isolation of active natural products.

Farnsworth: No. I will tell you why. If you take all the local anaesthetics that have been derived from the structure of cocaine, it would be a broad array, because one pharmaceutical company may develop one analogue, then another company adds a methyl group and markets the second product. It would be very difficult to trace that. The same situation exists with analogues of atropine or hyoscyamine.

Mander: I appreciate that it would be very difficult, but we should bear in mind that the discovery of a lead compound is often just the beginning of a very large programme. So saying that only 119 active natural products have been found is misleading in terms of the benefit of the activity.

McDonald: In the agrochemical industry, the number of commercial products that have been derived from natural products is quite small; only 14–15 are in significant commercial use, and of those probably only about half a dozen come from higher plants as opposed to microorganisms. However, there is a significantly larger number of synthetic products sold for agrochemical use which are definitely derived from natural product leads, for example the synthetic photostable pyrethroids are related to natural pyrethrins and the carbamate insecticides can be traced back to inhibitors of acetylcholinesterase that were originally discovered as natural products.

Many natural products which are found to have a desirable biological effect also carry undesirable side effects. It is often possible by synthetic modification to arrive at a more specific effect. Going back to those statistics for the NCI Programme, which took a lot of compounds and whittled them down to zero, I am sure compounds were discovered there with pharmacological actions that could be used as tools in subsequent screening programmes to find improved chemicals.

Farnsworth: Of all the pharmaceutical companies I know in the US that are interested in research on plant materials—and that's probably only four—none is interested in discovering active chemical structures which will be patented and marketed as drugs *per se*. They are looking for lead structures from which they would prepare analogues.

Years ago in the old NCI programme, Dr Jonathan Hartwell was in charge, and I said to him one day, 'You collect thousands of plants from all over the world which costs NCI a fortune. You get a kilo of each plant material, make an ethanolic extract, dry it and you get 60 g. You send this to a contract laboratory and they use 5 g to test against a battery of tumours. What do you do with the 55 g left?' He said 'We throw it away'. So I asked him to send these extracts to me: I have in my warehouse 50 000 extracts from the old NCI programme.

I made some of these extracts available to a pharmaceutical company. They isolated a compound from one of the extracts that was stable for less than 24 hours. It was a novel hydroxylated polyacetylene acid that had some interesting biological activity. This compound was isolated from an extract that was more than 25 years old.

I asked Dr Hartwell, 'What happens if a technician injects one of these extracts into mice and they all fall asleep for 12 hours; does anybody record that?' He said: 'Our goal is development of anticancer drugs. We would just test lower concentrations until the mice didn't fall asleep, but there would never be any record kept of that.'

The new NCI programme is all implemented under government control at Frederick, Maryland. Plant material is kept under deep freeze conditions; extracts are prepared and kept at $-70\,°C$. Theoretically, if you wanted to request small amounts of selected extracts, reasonable requests should be granted. This is a marked improvement over the old NCI programme.

McDonald: You mentioned the observation of a compound which was clearly photolabile and would go off on the bench top. It's our experience that a lot of natural products found to have biological activity also have a limited robustness in some way; it may be photoinstability, it may be instability towards hydrolysis or photooxidation. Again, synthetic chemists have succeeded in taking the original compound and building in the physicochemical properties which are needed to make the compound of more value to mankind.

About the storage of extracts, I would love to believe that we could get the best possible value from the enormous efforts and resources that go into the extraction of natural products, but it's important to bear in mind that secondary transformations will be going on during storage. Using old extracts to search for activities could be really frustrating, in that one may discover activities which are very difficult to reproduce because secondary chemical reactions have occurred.

Fowler: My concern is the way in which extracts are made and stored. I am sure that everyone in this room makes plant extracts in a different way, which will have some bearing on the results of bioassays. We have experienced major enzymic degradation within extracts shortly after preparation.

Concerning storage, do you believe there ought to be a standardized technique for extraction of plant material to be included in your database?

Farnsworth: No. Whenever you impose standards on scientists they usually object. If you screened 100 plants for anti-HIV activity and you made a water-soluble extract, you would probably get about 40 active hits because any sulphated polysaccharide will inhibit the AIDS virus but will never be useful in humans because these compounds are not active orally, they affect blood clotting mechanisms and so forth. Therefore you initially would have to extract with methanol, in which the polysaccharide is not soluble, and then do your fractionation. You could hold back some material and later, if you are interested in sulphated polysaccharides, you could look at those.

The extraction/fractionation scheme has to be developed on the basis of the bioassays you are going to use. But it doesn't take much material. We ask our collaborators to supply 100 g dry weight of plant material. We make an extract of 10 g and hold back 90 g. You would be surprised how many dozens of assays you can do with an extract from 10 g of plant material. The best standard would be a straight alcohol extract, if you had to have only one.

Fowler: Does your database detail the extraction technique used for each of the molecular species listed?

Farnsworth: Only to the extent that if you were interested in, for example, how to isolate azadirachtin from Neem, you could ask for all the occurrences where people have isolated it. But you would have to go back and look at the original papers for details. If you ask for the pharmacology of an extract of Neem fruits, the database would tell you whether it was tested as a hexane, ether, methanol, or other type of extract.

Reference

Noble RL, Beer CT, Cutts JH 1958 Role of chance observations in chemotherapy: *Vinca rosea*. Ann N Y Acad Sci 76:882–894

Ethnobotany and the identification of therapeutic agents from the rainforest

Michael J. Balick

Institute of Economic Botany, The New York Botanical Garden, Bronx, New York 10458, USA

Abstract. Many rainforest plant species, including trees and herbaceous plants, are employed as medicines by indigenous people. In much of the American tropics, locally harvested herbal medicines are used for a significant portion of the primary health care, in both rural and urban areas. An experienced *curandero* or herbal healer is familiar with those species with marked biological activity, which are often classified as 'powerful plants'. Examples are given from studies in progress since 1987 in Belize, Central America.

The Institute of Economic Botany of The New York Botanical Garden is collaborating with the National Cancer Institute in Bethesda, Maryland (USA) in the search for higher plants with anti-AIDS and anticancer activity. Several strategies are cited for identification of promising leads from among the *circa* 110 000 species of higher plants that are present in the neotropics, the focus of this search. Recommendations are offered for the design of future efforts to identify plant leads for pharmaceutical testing.

1990 Bioactive compounds from plants. Wiley, Chichester (Ciba Foundation Symposium 154) p 22–39

Of the 250 000 species of higher plants known to exist on earth, only a relative handful have been thoroughly studied for all aspects of their potential therapeutic value in medicine. Yet the plant kingdom has yielded 25% or more of the drugs used in prescription medicines today (Farnsworth 1988). This paper is focused on the unstudied portion of the plant kingdom, particularly on strategies for increasing the efficiency of the search for new therapeutic agents, given the rate at which natural vegetation, especially in the tropics, is being destroyed.

The New York Botanical Garden Institute of Economic Botany (IEB) began its research programme in 1981. One focus of this programme is the search for plant species with new applications in agriculture, industry and medicine. In October 1986, the National Cancer Institute (NCI) awarded the IEB a contract to collect 1500 plant samples from the neotropics annually for its anticancer and anti-AIDS screening programmes. Because the number of species of higher plants in the neotropics is estimated to be 110 000, we decided to test several approaches to determine which system of collection could generate the largest proportion of leads or 'hits' in the *in vitro* screens. By early 1990, IEB scientists or collaborating scientists from other institutions had collected plants in twelve countries as well as in the Commonwealth of Puerto Rico (Fig. 1).

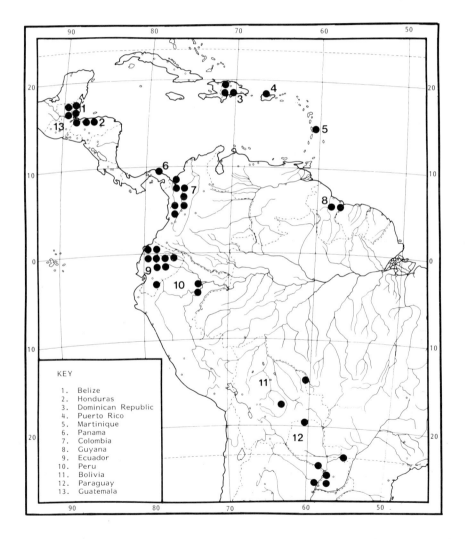

KEY

1. Belize
2. Honduras
3. Dominican Republic
4. Puerto Rico
5. Martinique
6. Panama
7. Colombia
8. Guyana
9. Ecuador
10. Peru
11. Bolivia
12. Paraguay
13. Guatemala

FIG. 1. Collection sites from 1986–1990 for the National Cancer Institute-sponsored plant collection programme.

Collection strategies

The random method

This method, as shown in Fig. 2, involves the complete collection of plants found in an area of tropical forest. In most cases, only plant species in fruit or flower are collected, as determination of sterile specimens can be time consuming, difficult, or occasionally impossible. Large numbers of species can be collected in this way, depending on the season and the number of fertile plants present in the area.

Targeted plant families

The second strategy is to target for collection those plant families known to be rich in biologically active compounds, such as alkaloids, glycosides, steroids or flavonoids. Fig. 3 illustrates an area of tropical forest but in this example plants from four botanical families are the focus of the collection because they are known to produce biologically active compounds: Apocynaceae, Euphorbiaceae, Menispermaceae and Solanaceae. Naturally, there are many other families rich in biologically active compounds.

RANDOM SAMPLES

FIG. 2.   Strategy for random collection in tropical forest. Diagrammatic aerial view of forest showing marked plot where plant species are collected.

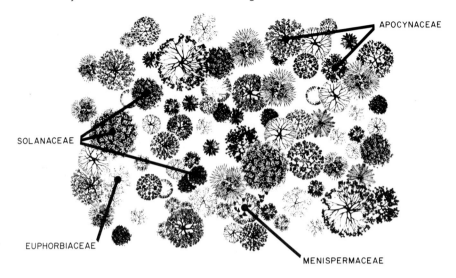

FIG. 3.   Strategy for plant collection in tropical forest by plant family on the basis of known presence of bioactive compounds in certain families.

The ethnobotanical approach

This approach employs local people's knowledge about the medicinal uses of the plants and their environment. These people, working with an ethnobotanist, select the species of plants that are used medicinally in the area. Fig. 4 depicts an aerial view of the same area of tropical forest as Fig. 3 in which four genera of plants are used for medicinal purposes: *Bursera*, *Protium*, *Simaruba* and *Strychnos*. This strategy is the most difficult and intellectually challenging one, as it involves identifying people who are knowledgeable about the medicinal uses of their flora, and securing their cooperation in identifying the species they use or know.

It is the ethnobotanical strategy that I will focus on in this paper. Most of my work for the NCI programme involves the ethnobotanical strategy, primarily working in Belize. The co-principal investigator on this project, Dr Douglas C. Daly, has carried out both random collections and ethnobotanical studies in Bolivia, Colombia, Ecuador, Martinique and Peru. Plants from other areas illustrated in Fig. 1 have been collected using the random approach by 23 collaborating botanists.

In July 1987, I travelled to Belize at the invitation of Dr Rosita Arvigo, a nature-cure physician and doctor of naprapathy (a traditional medicine regimen, using manipulation, diet and massage). An accomplished Western herbalist, Dr Arvigo had been studying with an elderly Maya *curandero* or herbal healer, Don Eligio Panti, for four years. Dr Arvigo wanted to collaborate with an ethnobotanist, to document the Maya pharmacopoeia, as utilized by

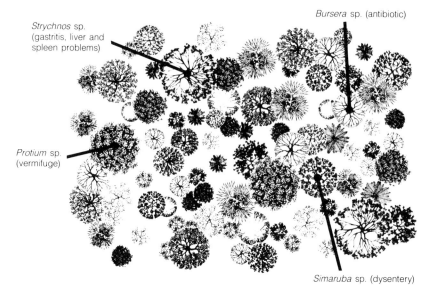

Strychnos sp. (gastritis, liver and spleen problems)

Bursera sp. (antibiotic)

Protium sp. (vermifuge)

Simaruba sp. (dysentery)

FIG. 4.   Strategy for collection in tropical forest based on ethnobotanical uses, in this case for medicines. Examples of useful plants taken from studies with the Maya in Belize.

Don Eligio Panti. After a few days with Don Eligio, and further discussions with Dr Arvigo, I recognized the opportunity to document botanically the *materia medica* of this *curandero* and agreed to return to Belize to work with Dr Arvigo, Dr Gregory Shropshire and Don Eligio Panti. This collaboration developed into the 'Belize Ethnobotany Project', a country-wide survey of the plants used by herbal healers, bushmen and other people knowledgeable about the uses of the native Belizean flora. Belize is a small country with some 180 000 people, but it is culturally diverse and the population includes Mopan Maya, Kekchí Maya, Mennonite, Ladino, Creole, Garifuna and North American groups. All of these people use plants to varying degrees, and comparative ethnobotanical studies are underway.

Traditionally, the major focus of ethnobotany has involved the preparation of lists of plants used by people. In the last few decades, an interdisciplinary approach has developed and ethnobotanists have joined with social scientists and researchers from other fields to gather information concerning useful plants (e.g. Berlin et al 1974, Davis & Yost 1983, Denevan & Padoch 1987, Vickers & Plowman 1984). These interesting collaborations have generated new perspectives on the study of the relationship between plants and people.

The ethno-directed sampling hypothesis

The ethnobotanical approach to collecting plants, as defined above, involves recognition of the activity as a process involving two components. The first component is the *cultural pre-screen* by local people of plants in their environment. This is usually thought of as a trial and error process occurring over hundreds or thousands of years. Another perspective, which is the one held by some of the *curanderos* we work with in Belize, is that the selection of useful plants is at least partially directed by supernatural forces. The second component is the *ethnobotanical filter*, in which the information about specific plant usage is acquired and introduced into the body of scientific knowledge through the ethnobotanical research. The *ethnobotanical filter* scans all the plants present in a region and captures those species of pharmacological use in order to, in the present case, provide samples to a drug development screening programme. The ethno-directed sampling hypothesis therefore maintains that the combination of indigenous knowledge about medicinal plants and the ethnobotanist's collection and documentation of this knowledge will yield a higher number of biologically active compounds for the screening programme on a per sample basis as compared to a group of plants collected at random. This will result in a higher probability of producing useful therapies from a group of medicinal plants. The pharmacological literature contains many references to the value of ethnobotany and indigenous knowledge in the development of therapeutic agents from plants (Cox et al 1989, Elisabetsky 1986, Farnsworth 1984, Svendsen & Scheffer 1982), although there are, to the best of my knowledge, no actual investigations carried out with a random collection as a control group. Later in this paper I shall discuss the data that we have received from the NCI's study in relation to the ethno-directed sampling hypothesis.

The *in vitro* anti-HIV screen

The first priority for the plants collected in the NCI sponsored programme is screening against the human immunodeficiency virus (HIV), the causative agent of AIDS. Plants are air or heat dried (below 65 °C) in the field and packaged in 0.5–1.0 kg samples, which are sent to The Frederick Cancer Research Facility Natural Products Repository in Frederick, Maryland for extraction and study. Voucher (herbarium) collections are made in the field and accompany the bulk samples to the US for distribution and study by botanists. Duplicate specimens are deposited in herbaria in the country where the plant was collected.

The dried plant samples are stored at $-20\,°C$ for a minimum of 48 hours immediately after arrival at the NCI. This period of freezing is a requirement of the US Department of Agriculture as a precaution to reduce risk of release of alien pests.

Each sample is labelled, either in the field or on arrival at the repository, with a bar-code label designating a unique NCI collection number. After freezing, samples are logged into a raw materials database and sent to the Natural Products Extraction Laboratory for grinding and extraction. A small portion of each sample is removed and kept as a voucher. The rest of the sample is then ground and extracted by slow percolation at room temperature with dichloromethane/methanol (1:1 mixture), followed by a methanol wash. The combined extract and wash are concentrated *in vacuo* and finally dried under high vacuum to give an organic extract. After the methanol wash, the residual plant material is extracted by percolation at room temperature with distilled water; lyophilization of the percolate gives the water extract. All extracts are returned to the natural products repository for storage at $-20\,^{\circ}$C until requested for testing.

In the *in vitro* anti-HIV screen, human T lymphoblastic cells infected with the AIDS virus are incubated for six days with varying concentrations of the extract (Weislow et al 1989). Untreated infected cells do not proliferate and die rapidly. Infected cells treated with extracts containing effective antiviral agents will proliferate and survive at moderate extract concentrations, whereas high concentrations of extracts generally will kill the cells. The degree of activity is measured by the level of protection provided by extracts at sub-toxic concentrations. Extracts are formulated for testing by dissolving in dimethyl sulphoxide and dilution with cell culture medium to a maximum concentration of 250 µg/ml.

Random versus ethnobotanical collection

To test the ethno-directed sampling hypothesis, two groups of plants were collected and sent to the NCI for screening against HIV. The random collection included plants from Honduras and Belize. The ethnobotanical collection was composed of species identified by Don Eligio Panti as 'powerful plants'. The 'powerful plants', according to Don Eligio, are species with substantial therapeutic value which are frequently used in his practice for a variety of purposes. The results from screening the two samples are presented in Table 1. It should be emphasized that these results are preliminary, and that data from only a small number of samples are available for analysis. However, these data have been analysed using the Fisher exact probability test, a methodology useful for handling data from small sample sizes (Siegel 1956). The results of this test show that $P=0.101$; while providing no firm proof, this is a strong indication that the null hypothesis (that there is no relationship between ethnobotanical

TABLE 1   **Preliminary testing of plants in an *in vitro* anti-HIV screen**

	Collection method	
	Random collections	*Ethnobotanical collections*
Total species tested	18	20
Per cent (number) active	6% (1)	25% (5)

collections and active compounds) should be rejected. Therefore, although the results from these two data sets do not conclusively prove the ethno-directed sampling hypothesis, they do support its validity. We expect that at the end of the five-year collection period, information on 7500 plant samples will be available, representing a minimum of 2000–4000 species, which can then be classified according to medicinal and non-medicinal uses and further analysed for their effectiveness in the anticancer and anti-AIDS screens. At that time we will also be able to detect trends by plant family, as discussed in the second collection strategy.

The conservation imperative and medicinal plants

Each year large areas of rainforest are destroyed through conversion to agricultural land or for other purposes. These habitats contain plant and animal species that are found nowhere else on earth. Most conservation activity focuses on the rainforest as an important and biologically rich habitat, and as a potential source of new products, such as medicines that could be introduced into commercial use. Rarely is the perspective taken that the forests contain a wealth of medicinal plants currently used by local people. It is estimated that up to 80% of the world's population uses plant medicines as an important component of primary health care (Farnsworth et al 1985) and certainly our fieldwork in Belize and elsewhere in Latin America supports this assertion. Therefore, since adequate hospitals and Western-trained doctors are not found in much of the tropics, the destruction of the rainforest will also destroy the primary health care network involving plants and traditional healers. Although no census has ever been taken, it is likely that there are thousands, if not tens of thousands, of traditional healers practising throughout the neotropics. Very often these people are elderly and, because of the forces of acculturation present in their society, many have no apprentices to carry on their work. One effect of deforestation is to reduce dramatically the supply of medicinal plants available to the traditional healer. Fig. 5 gives an idea of the situation facing Don Eligio Panti in the collection of plants he uses in his medicinal practice. In 1940, in an early stage of Don Eligio's practice, he walked for ten minutes from his house to the secondary forest site where he collected medicinal plants used to treat his patients. In 1984, when he began working with Dr Arvigo, it took them thirty minutes to reach a site that had sufficient quantities of these same species. In 1988, when I accompanied Dr Arvigo and Don Eligio Panti to the nearest collection site, it took us seventy-five minutes to reach an area where the plants could be found. Clearly, the added time and effort involved in the collection of medicinal plants is an increased burden for the traditional healer, therefore reducing his or her effectiveness in offering primary health care services to the community. Thus, in addition to building a case for conservation activities based on the *potential* use of the forest as a source of medicinal plants for our society,

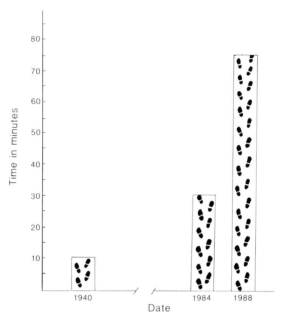

FIG. 5.   Time required for Don Eligio Panti to reach secondary forest site for medicinal plant collecting. Note substantial increase from 1984 to 1988, indicating high level of forest destruction in the San Antonio region of Belize.

there is also great strength in the contention that conservation of these species is essential to the survival of the existing health care network in many tropical countries involving traditional medicine. Otherwise, as effective therapies are identified and developed from tropical forest plants over the next 5–10 years, the original habitats in which the species are found may no longer exist, and the plants so desperately needed as a supply of raw material will have become extinct.

Conclusion

As tropical forests are destroyed and tribal peoples acculturated, our ability to discover new pharmaceutical agents and bring them into everyday use is being seriously compromised. The lull in natural products research over the last few decades combined with the reduction in global plant biodiversity has resulted in an urgent race against time. Given the relatively small number of scientists qualified to address this problem, it is clear that choices in the allocation of time and resources must be made. The ethno-directed sampling methodology allows the researcher to obtain a higher number of leads in a pool of plant samples as compared to a group of plants selected at random. The broader utilization of this methodology could streamline the discovery and development of drugs from natural products.

Acknowledgements

I would like to acknowledge the input of Drs Elaine Elisabetsky and Douglas C. Daly in developing the terminology for the ethno-directed sampling hypothesis and to thank Dr Gordon Cragg for the very clear explanation of the screening programme carried out by the NCI. Dr Elaine Elisabetsky and Wil de Jong kindly provided help with the statistical analysis and Dr Charlotte Gyllenhaal provided bibliographic assistance. Drs Brian Boom, Elaine Elisabetsky and Douglas Daly offered comments on an earlier version of this paper. Bobbi Angell and Carol Gracie prepared the graphics presented in this paper. The fieldwork in Belize and Honduras was sponsored by the National Cancer Institute, the Metropolitan Life Foundation, and the US Agency for International Development. I am most grateful to the Philecology Trust for support of my research. Active collaboration of the Belize Forestry Department and Ix Chel Tropical Research Center, San Ignacio, Belize and the Fundacion Hondurena de Investigacion Agricola, San Pedro Sula, Honduras is acknowledged. Finally, thanks are due Dr Rosita Arvigo, Don Eligio Panti and Dr Gregory Shropshire for their interest in and continued collaboration on the Belize Ethnobotany Project. It is Don Eligio Panti's fondest wish that we will find a way to build a bridge between the medicinal plant knowledge of the ancient Maya and the therapeutic needs of Western medicine and contemporary society, and we are working toward this. This small contribution is dedicated to those thousands of traditional healers in the neotropics who, for the most part, toil under the most difficult circumstances and in the face of an onslaught of 'modern' culture.

References

Berlin B, Breedlove DE, Raven PH 1974 Principles of Tzeltal plant classification. Academic Press, New York

Cox PA, Sperry LR, Tumonien M, Bohlin L 1989 Pharmacological activity of the Samoan ethnopharmacopoeia. Econ Bot 43:487–497

Davis EW, Yost JA 1983 The ethnomedicine of the Waorani of Amazonian Ecuador. J Ethnopharmacol 9:273–298

Denevan WM, Padoch C (eds) 1987 Swidden-fallow agroforestry in the Peruvian Amazon. New York Botanical Garden, New York (Advances in Economic Botany vol 7)

Elisabetsky EE 1986 New directions in ethnopharmacology. J Ethnobiol 6:121–128

Farnsworth NR 1984 How can the well be dry when it is filled with water? Econ Bot 38:4–13

Farnsworth NR 1988 Screening plants for new medicines. In: Wilson EO (ed) Biodiversity. National Academy Press, Washington, DC, p 83–97

Farnsworth NR, Akerele O, Bingel AS, Soejarto DD, Guo ZG 1985 Medicinal plants in therapy. Bull WHO 63:965–981

Siegel S 1956 Non parametric statistics for the behavioral sciences. McGraw-Hill, New York

Svendsen AB, Scheffer JJC 1982 Natural products in therapy. Prospects, goals and means in modern research. Pharm Weekbl (Scientific Edition) 4:93–103

Vickers WT, Plowman T 1984 Useful plants of the Siona and Secoya Indians of Eastern Ecuador. Fieldiana Botany, New Series 15. Field Museum of Natural History, Chicago

Weislow OS, Kiser R, Fine DL, Bader J, Shoemaker RH, Boyd MR 1989 New soluble formazan assay for HIV-1 cytopathic effects: application to high-flux screening of synthetic and natural products for AIDS-antiviral activity. J Natl Cancer Inst 81:577–586

DISCUSSION

Hall: Do you have a trained Western physician participating in your studies? A large proportion of people who come to Western doctors (I think some of my friends would say 60–70% of their patients) basically come for reassurance. So it's hard to tell which treatments are cures and which are not cures.

Balick: I have just led a scientific tour to Belize. In the group were two doctors, two nurses and five other lay people who wanted an experience in the rainforest and wished to help in the NCI plant collection programme. Some of these people were initially quite sceptical. At the end of the tour they admitted that they came wanting to see that a healer was prescribing a certain plant for something like melanoma and claiming to cure the disease—and they were going to disprove it. But they had realized that this was not the point; many patients come to herbal healers for help with primary health care. The Western-style hospitals are often too far away and too expensive for local people. Many of these people live to a ripe old age using traditional medical systems.

I would also say that these healers do not claim to cure all cancers or AIDS or some of the diseases that many of the MDs would try to pin them down on. They are curing diseases that are much more important to the people of the country—diarrhoea, malaria, fevers, dehydration, things that cause many deaths in Third World countries.

Kuć: I think it would be absolutely imperative to have medical doctors on the plant and information-gathering expeditions. It would be interesting and important to compare the diagnoses of the healers and the Western-trained medics, granted that the latter would not have the facilities they were accustomed to in making their diagnoses. The people that come to the healer mention a particular ailment. The healer then gives them, as I understand it, a particular drug for that illness. It would be very important to have documentation of what the healer was giving for particular ailments and what the medical doctor diagnosed. When the plant material is brought back to the West, one can relate the testing of that material to a particular illness. That might be more useful than the current approach where there are preconceptions of what you are looking for—a cure for AIDS, a cure for cancer—regardless of what the healer has in mind when giving the treatment.

Balick: That's a very good point. The patient interviews that I do take very careful note of the complaints, of the herbs administered and the quantity administered. We are also working with local Belizian doctors, so-called Western trained doctors and nurses.

I would like to find some MDs who would be willing to come with me. The conditions are not always very pleasant; some of the Western people I meet in the field do nothing but complain. There is a real opportunity here for people who are willing to get out of their own environment for a while, enter another environment and get some interesting data.

Cox: In my own research in the South Pacific we've had MDs participating. We find that MDs sometimes lack the anthropological ability to relate well to the people. I have had some MDs who have done very well, others who have experienced severe 'culture shock' and climbed back on the plane in two days. I find that sometimes physicians are culturally biased, so if people don't wear suits and ties and talk English and act in a Western manner, they are very uncomfortable in dealing with that other person, not only as a source of information but as a potential colleague.

My experience in the South Pacific, particularly in Samoa, is that the MDs I have taken down have been very impressed with what the people have been doing. The problem is trying to relate a specific remedy to a specific Western disease. Indigenous peoples do not use disease terms like we do; there is not a one-to-one correspondence between their disease classifications and ours. For example, where Michael Balick works a prominent disease is *susto*, which really does not translate into Western terms. And yet we have this sort of thing in European countries. In England there is something called gripe that babies have (they seem a little grumpy); the treatment is to give substances rich in ethanol to the baby. In Sweden, broken heart, loss of a romantic partner, actually causes disease and is recognized as an illness.

It takes a very special sort of MD to be able to undergo the sort of rigours of field work that Michael and I experience in our work. We welcome MDs who can approach other cultures with an open mind.

It is important to see our work as ethnopharmacologists in relation to the long-term needs of science. Everybody is interested right now in AIDS; they want substances that show anti-HIV activity. Obviously we should respond to that, but if we do our work properly we will not just respond to HIV, which in ten years may not be an issue, but we will address the drug needs of the 22nd century. A hundred years from now traditional antibiotics may no longer be efficacious and we may need new sources, new types of antibiotics. If Michael does his work correctly and I do mine correctly, our work will continue to provide a font of natural products for diseases and needs that we haven't even realized yet.

Balick: Let me give an example of a negative interaction between the MD and the healer. A medical person whom I met in the field once complained that the healer's shirt was dirty and therefore he wouldn't take him seriously as a health professional. Something as insulting, in a cultural way, as a statement like that could absolutely destroy years of work building a relationship with a healer. You have to balance the sort of suspicion and the arrogance of our own systems with the goal of the project which is to document the customs of the healers, understand it and utilize it before these people die.

Hall: You are making major claims that if we don't go out and do these analyses then things will be lost forever. Yet when we really look at the range of plants that you are analysing, I think the answer is that a large number of your patients, as I mentioned with Western patients, are concerned with emotional diseases. The fact is that someone sitting in the middle of the Bronx or in the middle of Chicago with no forest or plants around can be 'cured' by a good psychologist. So if the healer gives the patient something and tells them it's good, then it can cure people. I agree that in terms of human treatment that's important, but it doesn't relate to the search for bioactive compounds. I think the distinction hasn't been made of changes in emotional feelings. We certainly recognize that many plants, including tobacco, have major effects on people's emotions. Most of the discussion of these issues has been focused on diseases and we are sitting waiting for new cures.

Certainly the antihelminthic and antipyretic activities are effective. Dr Cox commented that there is no Western equivalent to some of these diseases; that's very worrisome because we need a precise focus on what we are talking about; otherwise we have no idea whether there is a specific compound that will deal with the aetiological agent. I am concerned that you may be getting a very large database but it won't contain the vital information about the purpose of a particular drug.

Balick: There is a notion that if something is not of physical origin it is not really a problem. I have a bias after living and working for fifteen years in environments where diseases are known to be caused by factors other than physical (e.g. infectious organisms). For example, I have seen someone go into convulsions after having their picture taken. They very strongly believed that their soul was being captured by the camera. It was a movie camera pointed at a person by a tourist without asking first. The person had to be hospitalized, they had to be treated by Western doctors, they had to be given valium and oxygen and a lot of other things and they were not released until three days later.

It is very hard for me to answer that kind of criticism because it assumes that these diseases or conditions do not exist unless they have been previously described by Western doctors. They do exist and that's where there is a very dangerous cultural gap between these two disciplines (ethnopharmacology and Western medicine). The only way we can bridge that gap is for you to come with me into the forest. Our surveys show that a significant number of patients come with problems caused by things like *susto*. Do you know what *susto* is? If something scared you and you went into a panic or a jaguar crossed your path, you could get physically ill from that. You would show symptoms of *susto*, which might manifest itself as a total lack of energy or 'spirit'.

The cultural heritage of a patient is very important. Perhaps we have to shed some of the burdens that we carry as scientists into the field and begin to consider things from other people's perspective, for example what are the disease concepts in a particular culture.

I witnessed a treatment for snake bite under the supervision of an MD in 1978 with the Guahibo Indians of the Llanos of Columbia. A man was brought into the hospital two days after being bitten by a very poisonous snake. He was in a state of toxic delirium, he had blood coming through the skin and blood in his urine. This patient was in the hospital and his condition was carefully monitored and documented. He was dying, he was given sufficient antivenom serum by a Western-trained MD to neutralize 200 mg of venom. It did nothing. A Guahibo *curandero* was there. He explained that the man did not understand what the needle was that the MD was putting into him. The *curandero* said 'You are not addressing his culture, you are not treating the spiritual side of his condition. Let me heal him, he's going to die, he has a couple of hours left, let me try.' The Western doctor was very open minded, having grown up and worked in Indian cultures, so he let the *curandero* treat the man. The *curandero* gave the patient a smoke-blowing treatment. He blew smoke and put water on the extremities of the patient. The patient was semi-conscious and was somewhat aware that this was happening. The results were amazing, the patient started to recover immediately: the symptoms started turning around, the heart beat changed, the blood, all of that started to improve. The doctor said that in the treatment of many snake bites he had never seen the patient's condition change for the positive so quickly. We wrote a paper together on this experience (Zethelius & Balick 1982). The great lesson is that if you don't treat the spirit, for people not accustomed to Western medicine, you are only going to be half as effective as you might be.

Elisabetsky: I would suggest that we are being too narrow by trying to find a single rule to deal with traditional information. Even in Western medicine, we treat each case individually. We have different pharmacokinetics, different dose ranges, routes of administration and so on. Now we are trying to find one rule to cover information from different cultures, each with their own traditional medical system.

We are trying to find plant species that are sources of cross-culturally effective drugs, so it does make sense to look for plants that are used in different cultures. What is being questioned here is the pattern of use. For instance, *Jatropha curcas* is used by 105 different cultures for 112 different purposes. That might be a sign of general toxicity rather than specific pharmacological activity. On the other hand the same disease can be expressed by different people in different manners. One people may use a plant to treat sneezing and another people may use it for itching, but we are talking about antihistamine activity in both cases. A plant may be used to treat allergy or inflammation and we can be talking about anti-platelet activating factor activity. So it is crucial to analyse the symptoms. We might not care to know that it's caused by susto but an epileptic fit can be described and interpreted as epilepsy. So even if MDs cannot go to the field (I had the same experience as Mike and Paul, it's very difficult to find an MD who is culturally prepared to respect a traditional healer), it's crucial to collect this information on symptoms and then work systematically.

Fowler: Am I right in thinking that a number of plants are used by different tribes and have different pharmacological effects in those tribes? In other words, the same species is used for different medicinal purposes.

Balick: No, one of the interesting things about studying five cultures and working with different healers in these cultures is that you also find examples of plants that are used to treat the same disease in different cultures. Remember that in the NCI programme, one of the cultural groups under study came from India, one came from Africa. Then there are other plants that have different uses in the various cultures.

Fowler: So to what extent are you picking up these variations because of a difference in the population physiology? And how would that relate to Western European or North American populations. We have a medicinal/clinical effect which may bear no relation to the Western situation, if there is a difference in the genotype of the human species.

Balick: In the case of parasites, in the case of stomach ache, eating bad food and feeling very sick, I don't have the medical training to say if there's a genetic difference between one group of people that gets *Entamoeba histolytica* and has one reaction and another that does not.

Farnsworth: I am intrigued by your finding that the plants selected on the basis of their use in traditional medicine have a hit rate against HIV considerably higher than that of randomly selected plants—I find this extraordinary. I believe your selections from traditional medicine were too small or they represented a specific plant family which ultimately will be shown to contain a certain widespread class of chemical compounds that will have biased your data.

Chapuis et al (1988) tested 200 extracts, prepared from 75 medicinal plant species used in Panama, Africa and Mauritius, for cytotoxicity. Fifteen of the 75 species showed activity in the single assay employed. The authors stated that pre-selection of plants on the basis of medicinal use might be a useful criterion for the identification of new bioactive compounds. There was no comparison of their pre-selected sample with a randomly collected sample. From these results, I find it difficult to rationalize the relationships between use in traditional medicine, cytotoxicity and new chemical compounds!

Balick: I don't know of any data sets that have compared random collections to targeted ones. Ours was a very preliminary attempt to examine the data we have from the NCI and present it. It is preliminary; I agree that 38 species is a small number. But those plants were from different families, from a diverse botanical spectrum. At present our analysis shows that this data set is borderline with regard to statistical significance. We will have a much larger data set three or four years from now when we have the other 7000 species analysed on our computers.

Of the 250 plants in the Mayan pharmacopoeia that we have identified, the ones that pop up over and over and over again in the healing process are the group that we refer to as 'powerful plants'. They run along certain curative lines, for example some are antihelminthics. By looking at the powerful plants that have very effective bioactivity, we can select species that will give higher numbers of hits in the screens.

Fleet: Would your HIV screen distinguish between general cytotoxicity and a specific antiviral effect? In this particular piece of work, you might just be looking at a set of plants that contain cytotoxic materials.

Balick: What I am suggesting is that by using this system we can maximize the number of leads that are found in the screens in a shorter time than with the random collections. I think we have shown that in a preliminary way. I cannot predict how many useful compounds are going to come out of this programme, if any. I view my job as the person in the field attempting to maximize the efficiency of your laboratory's time by delivering plants that will result in as many leads as possible.

I want to clarify the concept of the 'powerful plant'. I am not saying that if we are looking for a hepatitis cure we should go and look at all the healers who are treating hepatitis and then take only those plants back. I am suggesting that we look at the plants with the greatest amount of biological activity as identified by their frequent use by healers, and take those species back to the laboratory. That's how I got the anti-HIV data. I was not trying to bias the healer in any way, I simply observed and selected the most 'powerful plants' and provided them to the lab scientists. The result, admittedly very preliminary, is a fourfold increase over the random collections.

Bowers: I like the concept of the powerful plant very much. Certain plants produce a few chemicals in relatively high concentrations. If these chemicals do have some effect in biological systems, they will be more readily discovered than a chemical in the extract of a plant which may contain thousands of compounds but at only low concentrations.

An interesting example is the use of *Piper* species. Several years ago I was in Hawaii and I went to the botanical garden. With their permission I collected some leaves, took them back and put them into our tests against insects, fungi or nematodes. One of these plants contained a number of compounds which had very good nematocidal activity and also antifungal activity. These were allypyrocatechol, its diacetate, chavicol, chavibetol and chavibetol acetate. Two of those were previously unreported. The plant was *Piper betle*. In the literature I found that people in South East Asia chew the leaves of this plant after eating a meal. There were no claims about curing disease or a relationship with medicine, only that it was refreshing somehow. One report from the Philippines said that the leaves were sometimes wrapped around powdered lime and chewed after a meal. I made the calcium salts of those compounds and they were one hundred times more active against nematodes and fungi in the more soluble form (Evans et al 1984).

About that time I got a New Year's greeting card from Professor Sharma at the National Chemical Laboratory in Poona. This card was reproduced from an Indian painting of about 1750 AD. The fine print which explained the picture was a poem taken from this very old painting. The poem went 'clad in delicate fabrics of yellow, red and orange hue, the divine lovers seated on a low throne are enjoying the taste of betel leaf. (It was implied that this was *Piper betle*.) The handmaiden in the lower left corner is preparing betel from a gilded box. The sign of ecstasy on the lovers' faces is in tune with the hush of eventide that pervades the setting.' This was a very curious circumstance. Here is a plant the leaves of which have entered into the common use of people in many parts of Asia, just a cultural use, not a medicinal one. Yet there is sufficient of these compounds in one or two leaves to be effective against stomach nematodes and opportunistic fungi that might contaminate food. Chewing the leaves would be beneficial to health. So here is a link between cultural practice, presumably to refresh one's taste, and a clear medicinal utility associated with this sophisticated group of plants.

I looked in other species of *Piper* and found compounds that interfered with insect growth in various specific ways without being particularly toxic. Chemically, it is a very interesting genus. So I very much support the concept of powerful plants. These compounds are present in the plants at high enough concentrations that they can be detected in many assays.

McDonald: It is still worrying that maybe what you have there is a group of plants containing compounds that are relatively non-specific in their biological action, for example compounds that uncouple oxidative phosphorylation and interfere with respiration. In screening synthetic chemicals for biological activity one does come across these sorts of agents, which give an effect but are taken out of our current protocols for selecting compounds for clinical testing because of that very property. This is worrying in two ways: one is that perhaps from your natural sources these compounds will be less useful to medicine; the other is that modern pharmaceutical practice rejects compounds at an early stage which might be useful in medicine.

Cox: I would like to take up Norman Farnsworth's challenge. He asks why we should expect to find plants from Belize that are active against HIV? This virus wasn't found in Belize before about five years ago. There's no reason to believe that the people had that as a traditional pre-European contact disease category. There's no reason to believe that the people had specific remedies for HIV, so his challenge is quite cogent. Why should Michael Balick expect a higher rate of anti-HIV activity in plants derived from the healers in Belize than in plants collected at random?

To me an analogous situation would be the search for a bioactive molecule useful in treating another specific disease. Would we be more likely to find a useful substance by going through a pharmacy, pulling out all the drugs in the pharmacy and testing them, or by going through Chemical Abstracts and pulling out molecules at random? We know that the drugs in the pharmacy show bioactivity in human systems. Furthermore, we know that certain crucial problems of toxicity are unlikely to arise from these compounds; for example, they are unlikely to be mitochondrial poisons. Michael is generating plants that contain compounds that are bioactive in human systems. The belief is that by looking at those compounds we are more likely to find useful bioactivity than by just going randomly through the forest collecting plants that have no evidence of bioactivity. I believe that this explains why Michael would find a higher anti-HIV hit rate in plants derived from medicinal usage.

Balick: In addition, it's interesting that all of these plants are consumed orally. There are no arrow poisons that we found effective, for example. I think that gives you a lead on some of the cytoxic effects.

I think we will have the answer in about five years. We will have thousands of collections, we will have thousands of bits of information. I have no cancer data, even though the first plants were collected two and a half years ago. I agree that these are very preliminary data, it is a small data set, but it's the only one around.

Rickards: Dr Balick, is there any information on the chemical analysis of the plants you have identified, particularly those you call 'powerful plants'?

Balick: I use the NAPRALERT system and I asked Norman Farnsworth for a series of print-outs—some of the powerful plants are discussed in the print-outs. For some of the plants there is quite a lot of information available, for others there is none.

References

Chapuis JC, Sordat B, Hostettman K 1988 Screening for cytotoxic activity of plants used in traditional medicine. J Ethnopharmacol 23:273–284

Evans PH, Bowers WS, Funk EJ 1984 Identification of fungicidal and nematocidal components in the leaves of Piper-betle (Piperaceae). J Agric Food Chem 32:1254–1256

Zethelius M, Balick MJ 1982 Modern medicine and shamanistic ritual: a case of positive synergistic response in the treatment of a snakebite. J Ethnopharmacol 5:181–185

Ethnopharmacology and the search for new drugs

Paul Alan Cox

Department of Botany and Range Science, Brigham Young University, Provo, Utah 84602, USA

Abstract. Bioactive molecules occur in plants as secondary metabolites and as defence mechanisms against predation, herbivores, fungal attack, microbial invasion and viral infection. Throughout the world indigenous peoples have discovered plants with pharmacological activity; many useful drugs such as vincristine, reserpine, quinine, and even aspirin, have their origins in indigenous ethnopharmacologies. Three features characterize indigenous cultures that are most likely to have discovered useful natural products: (1) possession of a conservative ethnomedical tradition with established cultural mechanisms for accurate transmission of ethnopharmacological knowledge; (2) residence in an area with a diverse flora; and (3) continuity of residence in the area over many generations. Ethnopharmacological data derived from such cultures can be considered as analogues of human bioassay data. Yet indigenous ethnomedical traditions are rapidly disappearing. To preserve indigenous discoveries and make them accessible and useful to Western medicine, collaboration is required between linguistically sophisticated ethnobotanists, pharmacognosists, natural product chemists and pharmacologists. Ethnopharmacologists must be aware of all three components of indigenous medical traditions (a cosmological foundation, a repertoire of crude natural products, and a health care delivery system) to facilitate transfer of useful compounds and information to Western medicine, as well as to protect the interests and cultural integrity of indigenous peoples.

1990 Bioactive compounds from plants. Wiley, Chichester (Ciba Foundation Symposium 154) p 40–55

Plant natural products and new drugs

Natural products derived from higher plants may contribute to the search for new drugs in three different ways: (1) by acting as new drugs that can be used in an unmodified state, e.g. vincristine from *Catharanthus roseus* (Apocynaceae); (2) by providing chemical 'building blocks' used to synthesize more complex compounds, e.g. diosgenin from *Dioscorea floribunda* (Dioscoreaceae) for the synthesis of oral contraceptives; and (3) by indicating new modes of pharmacological action that allow complete synthesis of novel analogues, e.g. synthetic analogues of reserpine from *Rauvolfia serpentina* (Apocynaceae).

Natural plant products are also used directly in the 'natural' pharmaceutical industry that is growing rapidly in Europe and North America, as well as in traditional herbal medicine programmes being incorporated into the primary health care systems of Mexico, the People's Republic of China, Nigeria and other developing countries.

The importance of plant-derived pharmaceuticals can be deduced from their prominence in the market place: Farnsworth's 1984 analysis of the National Prescription Audit of 1976 revealed that 25% of all prescriptions issued in the United States and Canada contained an active component derived or originally isolated from higher plants. The need to examine the plant kingdom for new pharmaceuticals is particularly urgent in light of the rapid deforestation of the tropics and the concurrent loss of biodiversity throughout the world.

Why do plants produce bioactive compounds? With the exception of cultivated drug plants such as kava, whose mixture of kavalactones (structurally similar to sesquiterpenical lactones) has been produced through artificial selection by many generations of Pacific islanders (Lebot 1990), the pharmacological activity of most plants must be regarded as fortuitous. Bioactive molecules occur in plants primarily as secondary metabolites and/or chemical defences against predation, herbivores, fungal attack, microbial invasion and viral infection.

Random and targeted screens of floras for bioactive molecules

Modern searches for plants with bioactive molecules are typically based on some sort of bioassay that will detect specific bioactivity within a crude plant extract. Should activity be found, isolation, purification and structural determination of the bioactive molecule are then pursued. Given limited finances for such searches, it is clear that not every one of the 250 000 different species of flowering plants and thousands of conifers, ferns and bryophyte species can be studied for useful bioactivity. How then to decide which plant species should be selected for testing?

Two different types of plant selection programmes have traditionally been attempted in the search for bioactive molecules from plants. The first general type of programme I characterize as 'random', while the second general programme can be termed 'targeted'.

In random plant selection programmes, plants are collected and screened without regard to their taxonomic affinities, ethnobotanical context or other intrinsic qualities. Many large scale screening programmes of the 1960s, funded by both the private sector and governmental agencies, initiated essentially random screens of floras. Collectors were dispatched to various parts of the world and instructed to collect as many different types of plants as possible. Although some of these random screens were designed to ensure adequate botanical rigor, stories still circulate among the botanical fraternity of ill-trained collectors, usually chemists, hastily scribbling field notes directly on the leaves of plants, grinding up voucher specimens, and making radical mis-identifications of plants collected. While most of these stories were undoubtedly apocryphal, many screening programmes obviously lacked botanical rigour. Yet even rigorous random searches had an extremely low success rate.

Perhaps the lack of interest of the pharmaceutical industry in natural plant products during the 1970s and early 1980s stemmed in part from the failure of random searches. An exception to the rather dismal success rate is Taxol, an antitumour compound developed by the National Cancer Institute from *Taxus brevifolia* (Taxaceae), which is found in British Columbia.

The alternative to random plant surveys is targeted surveys. These can be of several types. In phylogenetic surveys, the close relatives of plants known to produce useful compounds are collected and analysed for either novel homologues or increased concentrations of the compounds of interest. For example, vincristine and vinblastine occur in very low concentrations in *C. roseus*; analysis of the close relatives of the taxon may reveal the same or similar vinca alkaloids in higher concentrations. In ecological surveys, plants living in particular habitats or possessing certain types of growth habits are chosen for study. For example, I am very interested in plants occurring in areas where slug and snail predation of seedlings is high—my hope is to find an inexpensive molluscicidal compound that could aid in the fight against schistosomiasis in central Africa. Finally, in ethnopharmacological surveys, plants used by indigenous peoples in traditional medicine are chosen for study. I will address the remainder of my remarks to this final category.

Ethnopharmacological screens and the search for bioactive compounds

It is of interest that most of the drugs derived or originally isolated from higher plants were discovered in an ethnobotanical context. For example, reserpine, derived from *Rauvolfia*, was discovered by analysing Indian Aryuvedic remedies. This is not too startling since many natural products from European plants, such as scopolamine or digitalis, were originally discovered because of their use, in crude extracts, in folk medicine. Yet such discoveries have continued well into the 20th century, perhaps the most significant being the discovery of vinca alkaloids, vincristine and vinblastine, from *C. roseus*. This was the 40th of 200 plant species examined by Eli Lilly and Co. from a collection of medicinal plants used by the indigenous people of Madagascar (Tyler 1986).

The history of drug discovery and development seems to confirm that ethnobotanical screens of floras have a far higher chance of success than random screens. But not all cultures are equally likely to use plants with significant pharmacological activity. Three features characterize indigenous cultures that probably do use such plants.

First, the culture must possess a conservative ethnomedical tradition with established mechanisms for the accurate transmission of ethnopharmacological knowledge. Cultures that have been using the same medicinal plants for many generations are more likely to have identified plants with useful bioactivity.

Second, the culture should reside in an area with a diverse flora. The high plant diversity should increase the probability of discovering plants with useful bioactivity. Using this rule, the Alyut tribes of northern Alaska are unlikely to have identified as many bioactive plants as the indigenous cultures of the Amazon basin, all else being equal.

Third, there should be a continuity of residence in the area over many generations. This provides more time for cultural exploration of local floral resources. Under this rule, the use of indigenous plants by the European settlers of Australia would be of less pharmacological interest than that by aboriginal peoples who have lived there for 19 000 years.

Ethnopharmacological data derived from cultures possessing all three of these characteristics are, in a sense, analogous to human bioassay data, particularly if people have been dosing themselves with the same plants for many generations. In such situations, preparations with low efficacy or acute toxicity are likely to have been discarded over the years. Thus, I argue that the greatest probability of discovering useful bioactive compounds from plants lies in careful examination of the pharmacopoeias of cultures possessing the three characteristics I have outlined.

Empirical vindication of the efficacy of ethnopharmacological screens

Is there any non-anecdotal evidence that indigenous pharmacopoeias are likely to contain plants with bioactive molecules? A recent survey of the Samoan ethnopharmacopoeia revealed that 86% of the plant species used in Samoan traditional herbal medicine show pharmacological activity in broad *in vitro* and *in vivo* screens (Cox et al 1989); 74 different species were tested for activity in a Hippocratic screen and a guinea pig ileum test. Further studies of Samoan medicinal plants have resulted in the isolation of a triterpene from *Alphitonia zizyphoides* (Rhamnaceae), which inhibits the formation of prostaglandin *in vitro* and EPP-induced oedema in the rat ear (P. Perera et al, unpublished) and antiviral compounds from *Homalanthus acuminatus* (Euphorbiaceae) (K. Gustaffson et al, unpublished). Similar results from studies of other ethnopharmacopoeias are being reported.

Unfortunately, though, few research teams have the necessary skills to initiate an ethnopharmacological screen. At the minimum such a team should have a core of two professionals: an ethnobotanist and a pharmacognosist. Pharmacology and botany have been historically disparate disciplines with differing approaches, and ethnobotany and pharmacognosy are the two disciplines most likely to bridge the gap between them in the search for novel pharmacological substances. Careful ethnopharmacological investigations by trained ethnobotanists are crucial not only in collecting the plants that will be later analysed, but also in documenting the methods of formulation, administration, and general cultural context of the plant use. Since the subsequent tasks of pharmacological screening and natural product chemistry are perhaps better understood by the pharmaceutical community, I will briefly describe the tasks of ethnobotanists in the search for new drugs.

Ethnobotanists and the search for bioactive molecules

Ethnobotanists are professionals with both botanical and anthropological training. Ideally, they should learn the language and customs of the people they work with, because it is their task to serve as bridges between different cultures, as facilitators of communication between the indigenous healers and the other scientists in the research team.

In the field, ethnobotanists are responsible for identifying the healers in an indigenous culture and securing their participation in the research programme. Such cooperation is usually granted, not only because of the language skills and cultural abilities of the ethnobotanist, but also because of the common knowledge and interest in plants shared by ethnobotanist and healer. Through extensive interviews, participant observation methodologies, and in some cases actual apprenticeships, ethnobotanists attempt to learn the ways of the healers and make their knowledge accessible to other Western scientists. This is rarely an easy task, since medicinal plants are only a single component of indigenous ethnomedicines.

The three components of ethnomedicine

All indigenous medical traditions share three components: a cosmological foundation, a repertoire of medicinal plants or other pharmacologically active substances, and a health care delivery system. An appreciation of all three components is necessary for rigorous ethnobotanical studies. For example, indigenous disease categories rarely can be mapped one-to-one on to Western disease categories; hence, understanding of the cosmological foundations of indigenous disease taxonomies and aetiologies is necessary to understand indigenous disease terms. Similarly, indigenous health-care delivery systems are usually very different from Western ones; fruitful cooperation with indigenous healers requires respect for local systems.

A rigorous ethnopharmacological research programme requires a high degree of cultural and linguistic sophistication on the part of the ethnobotanist. Rarely can adequate translations be obtained through interpreters, since healers employ specialist concepts and terms that are unknown to most of the members of their culture. As the anthropologist Bruce Biggs suggests:

'If you want to understand fully the ideas of sickness and health that underlie your healer's practices, his categories of disease, and the specialist vocabulary of

his profession, you must work in *his* language, not yours, because while you may, eventually, get to understand what he tells you in his language, and translate it into something that can be compared with western ideas on the same topic, there is no way that your informant, perfect though his English may be, can do that for you. His very use of English will mask and obscure your topic of investigation.'
(Biggs 1985, p 110–111)

Once rapport is established with the healers and a preliminary understanding of the cosmological foundation of the healing tradition is obtained, collection of the indigenous ethnopharmacopoeia can begin. It is crucial that any collections be documented with copious notes and well-prepared voucher specimens.

The importance of adequate herbarium voucher specimens cannot be overstated: any question or dispute concerning the botanical identity of the species involved can be unequivocally settled by examination of a properly collected voucher specimen. Such a specimen should supply the ethnobotanical information, including informant names, and geographical detail necessary to relocate the plant population should recollection be required (Croom 1983). Because of their scientific value for years to come (indeed, sophisticated bioassays of the future may require only micrograms of plant material), voucher specimens should be deposited and preserved in a well-curated herbarium with duplicate specimens deposited in geographically distant herbaria. The collection number of the voucher specimen should be used to label all subsequent pharmacological fractions and residues, so that any new discovery or question can be immediately referred to the original herbarium sheet.

In addition to preparing voucher specimens, ethnobotanists are also responsible for collecting plant materials for pharmacological testing. Careful attention should be paid to the plant parts collected, because frequently flowers, leaves, shoots and roots differ significantly in their chemical composition. Dried samples ranging from 0.5 to 2 kg have been traditionally supplied for testing, but drying, despite its ease, may not always be the best method of preservation. In general, collection and preservation techniques should closely approximate those used by the indigenous healers. Thus in Samoa, where healers prepare water-infusions from fresh plant materials (Cox 1990a,b) I macerate the fresh plants and pack them in 70% ethanol in shatter-proof aluminium bottles (Cox et al 1989). After transportation to the laboratory, the water fraction is freeze dried while the alcohol fraction is preserved in glass bottles for analysis by gas-phase chromatography.

Should any plants collected show useful bioactivity during initial screening, documentation should be sufficient to facilitate rapid recollection of material in bulk. Ethnobotanical data on the diseases treated, mode of formulation and methods of administration should be comprehensive, to guide subsequent investigations.

Obligations of ethnobotanists to indigenous peoples

The debt of the research team to the indigenous healers should be remembered during all phases of drug searches based on ethnopharmacological screens. By the time a drug search has advanced to the phase of organic synthesis, it is easy to forget that the compound in the test-tube was originally obtained through the kindness and willingness of traditional healers to share their knowledge. Because of her or his close association with the healers, the ethnobotanist in any research team has a special obligation to represent the interests of the indigenous healers. Thus it is crucial that the purpose and nature of the research programme be accurately represented to healers before one secures their participation. Second, during the ethnobotanical research every effort should be made to respect and affirm the indigenous culture. As a result, I do not solicit, record or transmit information that a healer identifies as 'secret'. I make it clear to the healer at the onset that I intend to publish the information she or he gives me. My experience is that healers are usually delighted to have their knowledge preserved. Wherever possible, I include a native language abstract in the publication, acknowledge the healers by name, and distribute reprints to the healers. Third, I make it clear to my collaborators that the indigenous healers have a financial interest in any marketable compound derived from their knowledge. These interests are written in patent agreements signed by representatives of the institutions and the villages concerned. In unmonetized cultures, financial interests can be translated into conservation efforts to preserve and protect rain forest and other habitats valued by the healers.

Conclusion

Although historically many useful bioactive compounds have been identified from indigenous pharmacopoeias, there have been few attempts to screen these carefully for bioactivity. Ethnopharmacological screens are far more likely to discover useful bioactive compounds than are random screens, but they require careful collaboration of ethnobotanists with pharmacognosists, natural product chemists, pharmacologists, toxicologists and other members of the research team.

References

Biggs B 1985 Contemporary healing practices in east Futuna. In: Parsons CDF (ed) Healing practices in the South Pacific. Institute for Polynesian Studies, Laie, Hawaii, p 108–128

Cox PA 1990a Samoan ethnopharmacology. In: Wagner H, Farnsworth NR (eds) Economic and medicinal plant research, vol 4: Plants and traditional medicine. Academic Press, London, p 123–139

Cox PA 1990b Polynesian herbal medicine. In: Cox PA, Banack SA (eds) Plants and man in Polynesia. Dioscorides Press, Portland, in press

Cox PA, Sperry LR, Tuominen M, Bohlin L 1989 Pharmacological activity of the Samoan ethnopharmacopoeia. Econ Bot 43:487–497

Croom EM 1983 Documenting and evaluating herbal remedies. Econ Bot 37:13–27

Farnsworth NR 1984 The role of medicinal plants in drug development. In: Krogsgaard-Larsen P, Christensen SB, Kofod H (eds) Natural products and drug development. Ballière, Tindall and Cox, London, p 8–98

Lebot V 1990 Kava (*Piper methysticum* Forst. f.): The Polynesian dispersal of an oceanian plant. In: Cox PA, Banack SA (eds) Plants and man in Polynesia. Dioscorides Press, Portland, in press

Tyler VE 1986 Plant drugs in the twenty-first century. Econ Bot 40:279–288

DISCUSSION

Bowers: You have done a lot of work in an island habitat: do the plants differ from those on the mainland?

Cox: Yes, plants on the islands do differ from those on the continents. There is not the range of insect predators and there is often a lack of mammalian herbivores. Endemism may, however, result in a high probability of novel compounds in island floras.

Bowers: Does this geographical isolation lead to novel chemistry?

Cox: Well, Anacardiaceae resins cause contact dermatitis in Samoa, as they do elsewhere. But founder effect through limited colonization and genetic drift in isolated plant populations can result in rapid speciation and hence novel chemistry.

Mander: I was puzzled by the comment that you were trying to duplicate the methods used by the healers by using plant extracts made in 70% alcohol.

Cox: The Samoan healers usually use an aqueous extract. They macerate the fresh plant material, extract in water and then rub that water extract together with fresh plant material directly on the skin of the inflicted person. I tried to figure out a way in which I could best duplicate or at least approximate that and avoid fungal or microbial contamination. My answer was to use 70% ethanol with the hope that the 30% aqueous portion would mimic what the healers were doing.

I didn't want to dry these substances, because the healers place a tremendous importance on fresh material. They do not store material, they wait until a sick person comes into their hut, make a diagnosis and then they go into the forest and find what they need. Perhaps there was a volatile compound that was efficacious, which drying might distort or lose. This was the best idea that I could come up with but I will entertain any suggestions on how I can better do this. As an ethnobotanist, my job is to present to you as a pharmacologist or natural product chemist essentially what the healers are giving to the people, so that you can make an accurate test of it.

Fowler: You may find that with 70% ethanol extraction, you get a certain level of polysaccharide precipitation and binding of the active component to that polysaccharide. We have observed this with some alkaloids.

Kuć: If the healer is applying material in an aqueous extract or suspension freshly prepared, the active components could be proteins, simple or complex carbohydrates, or lipids. The alcoholic preparation would denature the protein and you may lose activity; a 70% alcoholic extract would not dissolve complex carbohydrates or strongly lipophilic substances.

Bowers: If this is being applied topically, all the protein in the world would have no medicinal value, unless you were putting it on an open wound or abrasion. If you are making an extract that really has some ability to penetrate and has some medicinal activity, it must be an organosoluble compound.

Cox: It's interesting in that regard that usually the healer rubs coconut oil on the body of the person before they put the extract on. I don't know if there are lipophilic compounds present. The interesting thing to me is their reason for rubbing the macerated plant on the skin. The skin is an important site of administration since in their view there is a 'window of toxicity'. They believe that if they don't give enough of the compound it won't do any good, if they give too much there will be problems of toxicity. They think that putting it on the skin is a way of giving a slow controlled dose. Some of my physician friends feel this is a perfectly sensible way of administering certain substances.

Battersby: In the preparation of calabash curare by native people, extracts are made of *Strychna toxifera* and sprayed onto bark to dry in the sun. Photochemical reactions that are now fully understood increase the efficacy by 100–200 times.

Balick: I was very pleased that you presented the notion of indigenous rights and the importance of giving back something to these people and thereby putting a real value on it. In our work we give a copy of the data directly back to the people from whom we receive the plants and it is always remarkable to see the sense of pride that this information instils in the people. They feel that the plant has been 'certified' by that scientist. We sit with them and have discussions all night about the data, even before it's published.

Cox: I have a solar-powered computer and a battery-powered printer. In the field, each night I print out my notes on what the healer told me, including all the plants that they have suggested. I give a copy to the healer. They are pleased to see their words in writing. More importantly, they read through it and will correct me if I made a mistake.

Farnsworth: The World Health Organization's global programme on AIDS has a policy now that if you get information from a traditional healer that eventually ends up as a marketable product there has to be a statement, before they will fund a project, about how the informant and their community will share in those benefits.

Cox: I think that's terrific and I also think that the country should benefit. Some tropical countries, although very poor, may be sitting on a storehouse of bioactive molecules. I think it's crucial that it isn't just European countries or the United States that benefit economically. If there are economic benefits, they should benefit the country of origin as well. I am very pleased that WHO agrees with that concept.

Potrykus: I would like to ask a question about the ethnopharmacological and ethnobotanical approach in general. I have been trained as a natural scientist and I have learned that it is advisable to have controls in your experiments. I know from my own experience that not running controls can easily get one into problems, and I was trying to think about controls built into these approaches. The naive question I have is: has it been tested whether these extracts which are used by the healers work in the hands of non-healers? The second question is: have the extracts that you are preparing and sending to companies for tests been tested by the healers and have they been shown to be effective in the hands of the healers?

Cox: It is very important to realize that the work we do as ethnobotanists and ethnopharmacologists is not experimental work, it is observation and description. Our goal is to report accurately what the people do. If we were to do controlled work at the basic herbal level, in other words to do double-blind studies, have healers participate by giving some plants that they believe to be efficacious, others they believe not to be, or have extracts administered by healers and non-healers, we would run into a number of ethical difficulties. First, we are not MDs. Second, with government funding there are prohibitions against using human subjects in our research. My response, therefore, is we don't use controls because we are not doing experiments: we are making observations and describing what we see.

Elisabetsky: Dr Potrykus wants to know whether the ritual part of the treatment is important or not. I would say that in the Brazilian Amazon many of these drugs or remedies are used by mothers to treat their children and their husbands. So in many cases there is no ritual attached to the treatment. I agree with you that a less ritualistic kind of treatment might be more likely to reveal cross-culturally effective drugs.

Potrykus: I thought you wanted to show that your work is interesting because you are preserving important information. You were referring not only to the psychological effects of these treatments but also to the natural basis of these treatments.

Cox: Again, as an ethnobotanist I am trying to describe the indigenous culture. I am not making any claims about the efficacy of these herbal remedies. This is an important point that needs emphasis. As an ethnobotanist, sometimes people call me and ask 'Do you have something that can treat my brother's throat cancer?' or 'Do you have something for my sick child?' The answer is no, I am not an herbalist, I do not use plants to treat people. I direct such people to see their physician. I make no claim about the efficacy of indigenous herbal medicines. But my belief is that they indicate interesting plants to examine in a systematic fashion in Western science.

Returning to your more general point, the indigenous people are doing herbal medicine: I am not doing that, I am merely observing and reporting what they do. I don't see how a control procedure could be built into that. For example, if I were to study social interactions of people in Basle, Switzerland, how would I make a control? It is not an experiment, only an observation; but it still falls within the realm of science.

Potrykus: I am discussing this point because I could well imagine how my students would love to do exactly what you are doing, it's very popular at the moment. It is also popular to look down on the chemists and the pharmaceutical companies who are using chemistry to try to devise new drugs which are effective. I sense a strong trend in our students against exact natural sciences, against what has been done for the past fifty years and towards a 'back to Nature' approach. Because of this, I think it's important to discuss what is behind your observations. I think it is only fair to judge ethnopharmacological approaches by the same standards as traditional Western pharmacological approaches. This means that if one is excited about the potential of folk medicine, the folk medicine should be compared on the same standards as modern medicine.

Cox: Performing scientific research under difficult field conditions may be less attractive to your students than you imagine. Many years of linguistic and botanical training are required. But the point is that we are not trying to introduce folk medicine to Western countries, nor are we supplying plants for medicinal use. We seek to collect plants for pharmacological analysis, and isolation and structural determination of the active compounds. We are not, as you suggest, advocates of a 'back to Nature' approach. And ethnobotany, though not experimental, is still an observational science, just as anthropology, astronomy or geology are.

Hall: Ingo Potrykus has made the point that even if this is purely observational, it is still being communicated as potentially of interest to pharmaceutical companies and worthy of support, for example, from government funds. Therefore it should be subjected to at least a simplistic level of controls.

Once we have these materials there is a very simple control that can be done. Dr Cox said that these materials were collected in ways that it was hoped represented the natural way of collecting of the healers, for example in 70% ethanol. Dr Potrykus suggested to me that an appropriate test would be to give your extracts back to the healers to see if they work. Because if you are sending materials for pharmaceutical testing that don't even work in the hands of the healers, then we are a long way from making an advance.

Cox: Again I would have severe ethical problems giving anything to healers for them to use—that involves testing on human subjects. In the US, there is a tremendous regulatory procedure that has to be followed before we test any substance in human beings. There is no way that I, regardless of my funding source, would want to give substances to healers to do, in essence, experiments on humans.

But we do have substances that have come from my work that show interesting bioactivity. An NCI team and I have isolated an anti-HIV compound from a Samoan medicinal plant, for which a patent has been applied for.

Balick: I think a couple of issues are getting muddled here. There is this situation now that more and more people in Europe and the US want to have natural drugs and promote herbalism. The medical communities are getting very upset and the US and Canadian government regulatory agencies are starting to clamp down. That is one issue, but not the one we are talking about now. We are not doctors, we do not prescribe medicines, we do not practise traditional medicine ourselves.

The other issue is that there are at least 250 000 species of higher plants that exist on the planet, maybe even 500 000. Given that 50–66% will be destroyed in ten years, we are offering researchers a way to pre-select from that huge number a smaller number of plants to be worked on in laboratories. I view my work as a type of service; I go out there and get things for the chemists to work on and develop interesting products.

The issue of the control—I agree with that; in my paper (Balick, this volume), I presented a control group, which was random collection compared with the ethnobotanical approach. For that very small data set, which was a couple of years work with the anti-HIV screens, there was a fourfold higher hit rate from the plants collected by the ethnobotanical approach than for the randomly collected control.

Cox: In my own work, the original intention was to compare two samples, my collection of medicinal plants used by Samoan healers and a random collection. We used a broad Hippocratic *in vivo* screen and a guinea pig ileum *in vitro* screen, which are designed to test general pharmacological activity. 65 parameters were studied for effects lasting up to three days (Cox et al 1989, Malone & Robichaud 1962, Livingstone & Livingstone 1970). Together with Lars Bohlin in Uppsala we have completed preliminary tests on medicinal plants. We intend to analyse a random sample, but that work has not yet been completed.

Kuć: I think that is very unfortunate, because without those controls the data are far less convincing.

Farnsworth: One has to be very careful that results from screens do show what you think they show! We did some testing years ago that was prompted by a report that an extract of *Bixa orellana* showed dramatic tranquillizing or central nervous system depressing effects in mice. These effects were demonstrated by use of an 'activity cage'. Mice are allowed to move freely around the cage which has an electronic 'bar' at each end. Each time the mouse crosses this 'bar' it is recorded and used as an indicator of the mouse's activity. Drugs that cause the treated mice to score a low count are regarded as sedatives, those that lead to a high count as stimulants. After treatment with the *B. orellana* extract, the mice huddled in the middle of the cage and did not move about very much, thus giving a low count. The extract could therefore be described as a depressant. However, we eventually found, by intraperitoneal administration of extracts to mice, that this inhibition of movement was seen if the extracts were irritating to the mice—the mice avoided moving because this minimized the irritant effects, not because of a depressant effect on their central nervous system.

My question is, how many of the 'positive' results you obtained could be due to 'false positive' test results?

Cox: Some could have been, but I think the broad basis of the screen would eliminate many false positives.

Kuć: I share your concern about preserving the rainforests and the cultures of indigenous peoples. But I am not convinced that your method of selection, of looking at the plants that are being used by healers, is not excluding many plants that would have great benefit, especially since you are collecting plants that you hope will be useful against cancer, AIDS etc. It is not certain that the materials used by the herbalists have any activity against these diseases, or even have any activity at all.

Cox: All the natural product chemists are getting their material from somewhere—where? That's a botanical question: how do you choose which plants to study? We have offered here our best reasoning on a way to select which plants you study. Natural product chemists could spend many years isolating new and novel compounds from plants around their backyard. However, we want to help find plant *bioactive* compounds that will be useful in human systems. We believe the best way to do this is to examine carefully plants that people have been dosing themselves with for many generations. You have raised a very important point. It's unlikely that any culture, particularly one in an area with very diverse flora, has actually discovered all the useful plants in that flora. So we know we are missing some plants, but we think that we are giving you the best chance we can by directing your attention to these medicinal plants. We think that's a better way of doing things than just sending somebody out in a jeep and grabbing whatever is handy.

Balick: Please understand that when we go into the field—do you remember those three charts in my paper (Balick, this volume, Figs. 2–4), the random, the family and the targeted approach—we do not exclude the random collections, we get out there and get those plants also. It is how we analyse the data that will prove or disprove the point. So far the ethnobotanical approach seems to be more successful than the random approach in anti-HIV screens.

Farnsworth: All we are saying is we think this is the way it ought to be done. You get the information from human situations, which could be considered as a pre-screen. There are certain guidelines as to the credibility of the information that one collects. Then you put the plant extract through a bioassay system and if you obtain positive results, that's good. The proof is isolating the pure active compounds and developing them through the system, ultimately to clinical tests. We are saying that by following the human use in traditional medicine you will get to that endpoint faster than by random screening.

Bellus: This is an excellent discussion and I must say I agree with parts of the arguments presented by both sides! What I do not understand, however, is why the target effects are anticancer and anti-HIV activities. Nature didn't design a molecule to solve the problems of the human race. I would like to emphasize that Nature designed products in plants which are effective against insects and other pathogens. In the ethnobotanical investigations, one should not forget these effects and all local knowledge concerning these properties should be recorded.

Balick: We do record all the information we find; this is put on an herbarium label and attached to the specimen. So a hundred years from now someone could refer to this specimen and understand what all that information is for.

The reason that I am looking for anticancer and anti-HIV activity is because I have a contract from the National Cancer Institute to be the supplier for their research into neotropical plants. I would be delighted to collaborate with anyone who would like to study the tropical or temperate flora for any purpose whatsoever. This is our 'motherhood' issue. I don't go out and ask the tribal people what they use to treat cancer. We are looking for biologically active plants. This is a service; you can have access to it or ignore it. We would be pleased to expand our knowledge of the chemistry and bioactivity of tropical plants. This work is terribly important and there's only perhaps 10–20 years left to answer these questions about bioactivity that we are just beginning to address now.

Bowers: In my experience you rarely find something that you are not looking for. Generally speaking we have some fairly specific ideas of something we would like to find that would help in medicine or in agriculture. We have assays that will detect the chemicals produced by plants as natural defensive compounds and usually we are able to refine them and make them very sensitive. Yet, we have to have the plant resources to plug into those assays and some knowledge of prior experience from native cultures. The ethnobotanists who make their observations amongst these indigenous peoples on their use of plants as drugs need to have specific questions to put to the healers to try and refine the selection techniques.

Cox: We have a very thorough protocol for this. I recently supplied Schering Corporation with 40 ethnobotanical samples and they were accompanied by 127 pages of ethnobotanical information.

Bowers: Is this available generally to those of us who work on these products?

Cox: Yes, it is. At the American Pharmacognosy meeting in 1988 I offered to supply my complete repertoire of samples to anybody who would test them. The only thing I asked was that they were tested. It is remarkable to see the lack of interest among pharmaceutical firms in natural products in general and particularly in natural products from plants. I think that may change. I, and some of my colleagues too, have visited pharmaceutical firms throughout the United States and Europe but only a few show any interest. This really is a pity. We will continue doing our work regardless, because we are interested in indigenous peoples and ethnobotany as a science in and of itself.

Bowers: Would you make these samples available to people other than pharmaceutical companies?

Cox: Of course. I collaborate with a number of scientists, such as with Alice Clarke's group in Mississippi and with Lars Bohlin's group in Uppsala, Sweden. We are delighted to share our information. Our sense is that we are generating high quality samples but we are limited because we do not have the facilities to do bioassays.

Mander: Is this attitude of the pharmaceutical industry simply a matter of fashion or has it made a careful decision about whether plants are going to be a productive source of new therapeutic agents?

Farnsworth: My considered opinion is that the pharmaceutical industry is so short of leads that they are now diverting into the natural product area. They recognize that natural products have some unique structures and they want to take a look at them. Most companies don't want to get involved in working with crude extracts.

The only pharmaceutical companies that I know in the United States that are interested in natural products are Merck Sharp & Dohme, Eastman Pharmaceuticals, Smith, Kline & Beecham and Glaxo. There are several small start-up companies getting into this area either from a mass screening viewpoint or from looking to ethnopharmacology to provide leads. This is a new trend; five years ago there wasn't a single pharmaceutical company in the United States that had any interest in developing drugs from higher plants.

Balick: I think the trend is towards greater interest in ethnopharmacology and in natural products—it certainly couldn't get any worse. In the last two years I have been contacted by at least a dozen companies. I am very careful about who I collaborate with in industry. I ask that royalty agreements be provided so that we can compensate countries and colleges in these countries. I ask that data be provided to the country so they realize that the colonial era is over and they are no longer to be 'mined' as countries. I ask that they be offered the first chance at being the provider of the natural product. Then I ask that the costs of the collection programme be covered, and it's surprising to me how many companies refused. Several companies have come back to me a couple of years later saying that they tried the cheaper option but it doesn't work.

The head of research division of a major pharmaceutical company, when I asked him why companies were renewing their interest in natural products, said 'twenty years ago the then generation of economists came to us and said drop all this natural products stuff, synthetic chemistry is in, we will guarantee you ten leads'. Ten years later the next generation of economists came to that same research director and said 'this synthetic chemistry has got us nowhere—the rainforest is it, drop the synthetics and put some money into natural products and we will guarantee you twenty products in the next ten years'. It is pressure from the business managers and financial people to develop more drugs.

Bellus: One reason for the renewed interest of the pharmaceutical companies in natural products is that we now have fully automated *in vitro* tests for many essential enzymes. Many lead compounds are now being discovered in this way. It is not only fashion that is changing, the method of discovery is changing.

Potrykus: I personally find ethnobotany and ethnopharmacy very interesting in themselves; they should not need commercial justification. It would be a pity if you had to justify your work by promising to find something for pharmaceutical companies: I think you have better reasons to do your work. Also there is a better argument for preserving the tropical rainforest than promising to find something which may cure cancer or AIDS. I am afraid that it is even rather dangerous to promise to find such drugs, because you will probably not find what is expected and then after ten years people will not listen to your arguments for protecting the rainforest.

References

Balick MJ 1990 Ethnobotany and the search for therapeutic agents from the rainforest. In: Bioactive compounds from plants. Wiley, Chichester (Ciba Found Symp 154) p 22–39

Cox PA, Sperry LR, Tuominen M, Bohlin L 1989 Pharmacological activity of the Samoan ethnopharmacopoeia. Econ Bot 43:487–497

Livingstone E, Livingstone S 1970 Pharmacological experiments on isolated preparations, 2nd edn. Edinburgh University Press, Edinburgh

Malone MH, Robichaud RC 1962 A hippocratic screen for pure or crude drug materials. Lloydia (Cinci) 25:323–332

Some problems in the structural elucidation of fungal metabolites

W. Steglich, B. Steffan, T. Eizenhöfer, B. Fugmann, R. Herrmann and J.-D. Klamann

Institut für Organische Chemie und Biochemie der Universität Bonn, Gerhard-Domagk Straße 1, D-5300 Bonn 1, Federal Republic of Germany

Abstract. In the structural elucidation of natural products, problems may arise owing to extensive signal overlap and line broadening in the NMR spectra as well as unexpected chemical reactions. This is illustrated in the analysis of some novel compounds isolated from fruit bodies of toadstools and slime moulds, including infractopicrin, lascivol, retipolide A, arcyroxepin A and rubroflavin. In some cases, the problems could be solved by spectroscopic techniques such as two-dimensional INADEQUATE experiments, nuclear Overhauser enhancement measurements or X-ray structural analysis; in other cases, chemical transformations were necessary to resolve the structure.

1990 Bioactive compounds from plants. Wiley, Chichester (Ciba Foundation Symposium 154) p 56–65

The rapid development of powerful spectroscopic techniques has led to the widespread belief that the structural elucidation of natural products is nowadays a matter of mere routine. In this paper I would like to point out that the natural product chemist may still encounter a number of difficulties during structural studies. I will illustrate this with some examples from our work on fungal metabolites.

Crystallopicrin

Several toadstools contain bitter principles which protect them against the attack of insects and other fungivores (Steglich 1989a). Thus, the viscid caps of *Cortinarius crystallinus, C. croceo-coeruleus* and *C. vibratilis* contain the extremely bitter triterpenoid, crystallopicrin 1. This compound resisted all attempts at crystallization and its ¹H NMR spectrum shows severe signal overlap in the aliphatic region which obscures nearly all structural information. The same problem was encountered in the application of two-dimensional correlation spectroscopy (COSY) and related techniques. In contrast, the ¹³C NMR spectrum exhibits clear signals from all 30 carbon atoms. The availability of 150 mg of crystallopicrin enabled a two-dimensional INADEQUATE experiment to be performed (Turner 1982), from which the complete carbon–carbon connectivities were obtained. The presence of a cyclic ether moiety was excluded by a deuterium-induced differential isotope ^{13}C NMR experiment (Pfeffer et al 1979). Structure **1** could then be assigned to crystallopicrin.

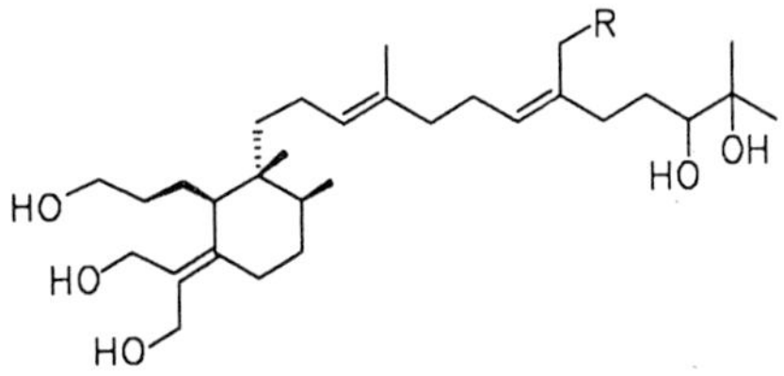

1, R = OH

2, R = H

The relative configuration in the cyclohexane ring was established by difference nuclear Overhauser enhancement (NOE) measurements. The absolute configuration of crystallopicrin has still to be determined. The compound is accompanied in *C. vibratilis* by its deoxy derivative **2**. Crystallopicrin is closely related to the iridals from *Iris* species, which have been investigated by Jaenicke & Marner (1986).

Lascivol

Another bitter substance, lascivol **3**, is present in the fruit bodies of *Tricholoma lascivum*. The fast atom bombardment spectrum and the elemental analysis indicate the molecular composition $C_{17}H_{28}N_2O_7$. The gross structure of lascivol could be discerned from the usual one- and two-dimensional NMR methods. It is supported by the results of a two-dimensional INADEQUATE experiment which also established the presence of a γ-glutamyl moiety. The determination of the relative stereochemistry of lascivol presented considerable problems and could only be solved by crystal X-ray structure analysis. Finally, the absolute configuration of lascivol was determined by acid hydrolysis to L-glutamic acid **5**. During this degradation 5-methoxy-2,4-dimethylindole **4** is formed from the amine part of the molecule by an interesting aromatization reaction.

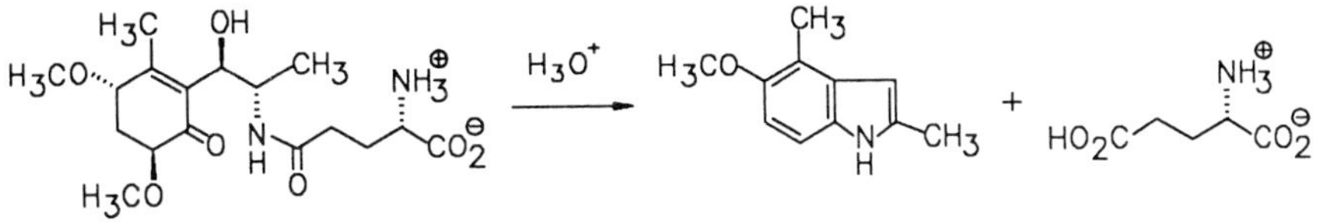

3   4   5

Retipolide A

Different problems are presented by retipolide A **6**, an antibiotic from the bitter-tasting fruit bodies of the North American bolete, *Boletus retipes*. The compound shows extensive line broadening of some signals in the ^{1}H and ^{13}C NMR spectra, which prevent the application of normal two-dimensional correlation techniques. Only the structure of the conformationally rigid macrocyclic lactone ring can be deduced from the spectral data. Ring strain leads to the magnetic non-equivalence of the aromatic protons in the 1,4-disubstituted phenyl group and this is reflected in the NMR spectrum. Strong shielding of 2-H by the neighbouring, orthogonal phenyl ring causes this aromatic proton to appear at δ 4.83. The structure of the 'heterocyclic' portion of the molecule was finally established by two-dimensional NMR experiments carried out with the isopropyl derivative **7**. In contrast to retipolide A **6**, this compound exhibits a complete set of sharp signals in the ^{1}H and ^{13}C NMR spectra. The line broadening of the signals belonging to the 'heterocyclic' part of **6** can therefore be explained by a rapid equilibrium between the hemiacetal and the ring-opened form. Structure **6** suggests a biosynthesis of retipolide A from three molecules of 4-hydroxyphenyl pyruvate. The latter is the biosynthetic precursor of several other bolete pigments (Gill & Steglich 1987).

6 , R = H
7 , R = iPr

Arcyroxepin A

Line broadening in the NMR spectra also led to difficulties in the structural elucidation of arcyroxepin A, a minor red pigment available in only milligram quantities from the slime mould *Arcyria denudata*. This tiny organism produces a fascinating mixture of closely related pigments, including arcyroxocin A **9**, which seem to be derived from the simple bisindolylmaleimide arcyriarubin A **8** (Gill & Steglich 1987, Steglich 1989b). The structure of **9** has been confirmed by an X-ray structure analysis. Pigment **9** is accompanied in *A. denudata* by a pigment with a similar R_f value in partition chromatography, which exhibits a characteristic colour change from red to violet on exposure of the thin layer chromatograms to ammonia. The mass spectrum of this compound indicates the molecular composition $C_{20}H_{11}N_3O_3$, which suggested that the new pigment is an isomer of **9**. The compound exhibits two broadened *ortho*-'doublets' and two overlapping 'triplets' in the aromatic region of its ^{1}H NMR spectrum ($[D_6]$ acetone) at ambient temperature. This implied a symmetrical structure in which the benzene rings of the indole moieties are unsubstituted. Because signals from the indole 2-H protons were apparently missing, we originally proposed structure **10** for this pigment and named it arcyroxepin A (Steglich et al 1980).

8   9   10

We became suspicious about the accuracy of this structure, however, when we observed that arcyroxepin A yielded arcyroxocin A **9** on prolonged standing in $[D_6]$ acetone. We subsequently discovered that this transformation can also be achieved by briefly heating arcyroxepin A in toluene to $110\,°C$. The problem was finally solved by methylation of arcyroxepin A with methyl iodide/K_2CO_3 in acetone, which produced a trimethyl derivative, $C_{23}H_{17}N_3O_4$, to which formula **12** could be unequivocally assigned by extensive NOE experiments. The structure of arcyroxepin A must therefore be corrected to **11**. This structure was confirmed by the quantitative reduction of this pigment to arcyroxocin A **9** on brief exposure to aqueous $TiCl_3$. Arcyroxepin A represents the first natural *N*-hydroxyindole derivative to be described (Somei & Kawasaki 1989). Compounds of this type are known to be rather unstable and easily reduced to the corresponding indoles (Acheson et al 1978). Furthermore, the potential existence of tautomers in solution explains the coalescence phenomena observed in the NMR spectra.

Rubroflavin

An unexpected chemical phenomenon also caused problems during the structural elucidation of rubroflavin **14**. This pigment occurs in the form of its leuco derivative **13** in the colourless fruit bodies of the North American puffball *Calvatia rubroflava* which turn bright orange when touched. In this case oxidase enzymes are apparently released that transform **13** into the quinone derivative, rubroflavin **14** (Gill & Steglich 1987). During our work on rubroflavin, we were initially puzzled by the fact that the electron impact mass spectrum indicated a molecular formula $C_8H_{10}S_2O_2$, whereas in the ^{13}C NMR spectrum nine carbon signals were clearly discernible. This discrepancy was eventually solved when we found that **14** undergoes a smooth thermal fragmentation involving loss of the semicarbazone moiety. During this reaction the phenol **15** is formed in nearly quantitative yield. Excellent agreement between the circular dichroism spectrum of **15** and that of (S)-methylphenyl sulphoxide (Mislow et al 1965) allowed us to assign the (S)-configuration to rubroflavin.

Conclusion

These examples demonstrate that the structural elucidation of natural products still offers surprises to the chemist and that, in those cases where the amount of material is limited, the structural elucidation and assignment of stereochemistry to more complicated compounds can lead to serious and challenging complications.

Acknowledgement

We thank the Deutsche Forschungsgemeinschaft for continued support.

References

Acheson RM, Hunt PG, Littlewood DM, Murrer BA, Rosenberg HE 1978 The synthesis, reactions, and spectra of 1-acetoxy-, 1-hydroxy-, and 1-methoxy-indoles. J Chem Soc Perkin Trans I, p 1117–1125

Gill M, Steglich W 1987 Pigments of Fungi (Macromycetes). Prog Chem Org Nat Prod 51:1–317

Jaenicke L, Marner F-J 1986 The irones and their precursors. Prog Chem Org Nat Prod 50:1–25

Mislow K, Green MM, Laur P, Melillo JT, Simmons T, Ternay AL Jr 1965 Absolute configuration and optical rotatory power of sulfoxides and sulfinate esters. J Am Chem Soc 87:1958–1976

Pfeffer PE, Valentine KM, Parrish FW 1979 Deuterium-induced differential isotope shift ^{13}C NMR. 1. Resonance reassignment of mono- and disaccharides. J Am Chem Soc 101:1265–1274

Somei M, Kawasaki T 1989 A new and simple synthesis of 1-hydroxyindole derivatives. Heterocycles (Tokyo) 29:1251–1254

Steglich W, Steffan B, Kopanski L, Eckhardt G 1980 Indolfarbstoffe aus Fruchtkörpern des Schleimpilzes *Arcyria denudata*. Angew Chem 92:463–464; Angew Chem Int Ed Engl 19:459–460

Steglich W 1989a Some chemical phenomena of mushrooms and toadstools. In: Schlunegger UP (ed) Biologically active molecules. Springer-Verlag, Berlin, Heidelberg

Steglich W 1989b Slime moulds (*Myxomycetes*) as a source of new biologically active metabolites. Pure Appl Chem 61:281–288

Turner DL 1982 Carbon-13 autocorrelation NMR using double-quantum coherence. J Magn Reson 49:175–178

DISCUSSION

Battersby: Can you give us an example where the elemental analysis was important?

Steglich: We have no facilities for doing field absorption studies. After we discovered that we had made this mistake in the structure of arcyroxepin A by relying on the electron impact spectrum, we sent some of the material to a colleague for testing. We saw absorption spectra typical of very small ions. If we had used this method at the beginning, we would have suspected our formula earlier.

Ley: Professor Steglich, you talked about the instability of retipolide A, the opening of the lactol in basic conditions. You didn't comment on the absolute configuration of the stereocentre. It seemed to me that you could lose that centre under those conditions, by a retro-Michael process, and then it wouldn't close back to the natural material. Is that what happens?

Steglich: We have only just arrived at this structure and we are now going to study the chemistry of this interesting compound. Retipolides A and B are optically active. There is, of course, the possibility of a Michael-type ring opening and re-closure of the γ-lactone, which would lead to racemization. However, the conformation of the 14-membered ring is so rigid that it may retain its stereochemical identity even after formation of the exocyclic double bond.

Ley: I was interested by these unstable *N*-hydroxyindoles. The oxygen is lost from the product very rapidly, do you know where it goes to? I believe it could act as an oxidizing agent for other substrates.

Steglich: We had the same idea, that this could be a very selective reagent if one had the right substitution at the indole ring, perhaps a carboxylate group in the 3 position. We have isolated only about 2–3 mg of this compound, so we have not done any follow-up of what happens to the toluene. It is probably oxidized to benzyl alcohol or something like that. *N*-Hydroxyindoles are so unstable that they are transformed into the corresponding indoles even on standing in suitable solvents.

Ley: We have made similar observations with sulphoxides and selenoxides, but were unable to determine where the oxygen goes to.

Ruchirawat: In your hydroxyindole system there are two indole moieties with very similar environments. How did you prove that the hydroxy group is attached to one particular nitrogen?

Steglich: The permethyl derivative gives an *O*-methyl signal in the ^{1}H NMR spectrum that is nicely separated from the *N*-methyl signals. This allowed us to locate the position of the hydroxy group without any doubt by nuclear Overhauser enhancement measurements.

Bellus: The arcyriarubins undergo many transformations catalysed by sunlight, e.g. 2 + 2-photodimerization. Did you find a photodimer?

Steglich: No we have not found this. It is a very nice set up for electrocyclic reactions. In our synthetic models we have tried to find a ring closure, perhaps a reversible one, but so far we have not succeeded. The arcyriarubins are very stable—we have some on thin-layer chromatograms which are now eight years old and they still have the same colour. So some of them may be potential dyes.

McDonald: Your rationale for the biosynthesis of crystallopicrin was fascinating. Have there been previous analogies for that type of skeleton or is your rationale based upon your thinking about the mechanism?

Steglich: We just thought about the mechanism. As I mentioned, Professor Jaenicke (Köln) came to the same conclusion when he studied the iridals from *Iris* species.

McDonald: The first ion that you showed was a sort of half-formed steroid, which has not been able to form a final ring. You also showed some speculations: the one I would like to ask you about most is the rubroflavin, where you have a simple six-membered ring with lots of sulphurs and nitrogens in unusual positions, particularly the two nitrogens linked together. Do you have any idea from the structures of related natural products in what order those heteroatoms are introduced?

Steglich: Basidiomycetes have a strong tendency to form compounds with N-N and N-O bonds, e.g. phenylhydrazine derivatives, formazanes, *N*-hydroxyureas and even diazonium salts. If you crush the fruit bodies of *Agaricus xanthoderma* under methanol and add some phloroglucinol or β-naphthol, red azo dyes are formed! The biosynthesis of agaritine, a glutamic acid (4-hydroxymethyl)phenylhydrazide, has been shown to start from 4-aminobenzoic acid which is diazotized and reduced to the hydrazine derivative. In the case of rubroflavin, oxidative decarboxylation of the carboxyl group may occur. The introduction of the two sulphur atoms is still a mystery.

Barz: Do you have any information on the localization of these compounds in the tissue? Are they in the outer layers?

Steglich: Several of these compounds are located in special compartments. For instance, the cap skin of the bay fungus (*Xerocomus ladius*) and related boletes contains unusual naphthalene derivatives which have a high complexing capacity for K^+. After the reactor accident at Chernobyl, this complexing ability led to the accumulation of ^{137}Cs in these mushrooms and there was a warning in Germany that you shouldn't eat them any more because of their high caesium content.

In *Lactarius* species, the defensive substances are located in the milky latex.

Barz: Are they really defence compounds of the lower plants, as you implied?

Steglich: Yes, some have very well documented properties in this respect. Dr Sterner and Professor Wickberg in Lund, Sweden, have investigated the defensive systems of *Lactarius* species. These toadstools have a white milk in special organelles which contains a kind of pro-drug, from which enzymic action releases two dialdehydes, isolvelleral and velleral, with potent antifeedant and antibacterial activities. After a few minutes, these compounds are detoxified by the action of a second enzyme. It is quite convincing because these mushrooms have a strong repellent activity that is due to the presence of this system. We have found other 'pro-drugs' in toadstools which may serve the same purpose.

Farnsworth: What kinds of yields do you get for these compounds, per weight of dried mushroom?

Steglich: In the case of crystallopicrin and lascivol, from 1kg of fresh mushrooms we can easily get 1–2 g of these compounds, which are present in amazing amounts. However, in other cases we obtain only a few milligrams. Working with myxomycetes is more difficult because they are so tiny. But we calculated that the material makes up 20% of these tiny fruit bodies, so even having only 200 mg, which you can collect in an afternoon, one can get several milligrams of compounds.

Mander: What are the prospects for growing these mushrooms under artificial conditions? If you find an interesting compound and you want to obtain larger quantities of it, how practical is it to culture these mushrooms?

Steglich: That is a major problem. Many of the basidiomycetes grow in mycelial cultures; however, they do not produce the same compounds as are found in the fruit bodies. Unfortunately, the general problem of how to induce fruit body formation from mycelial cultures is not yet solved. One day somebody will discover this miraculous compound and then we will be able to produce fruit bodies of all the desired fungi.

Yamada: Have you analysed the mycelia of your mushrooms? Do they contain the bitter component, crystallopicrin?

Steglich: Unfortunately, *Cortinarius* species from this group are very difficult to grow. Professor Anke in the biotechnology department of Kaiserslautern University has tried to grow this particular mushroom, but not succeeded.

Barz: It would be a great advantage to grow them in culture but an alternative would be to grow them under controlled conditions. In Japan, there is plenty of experience of growing higher fungi on hay or straw; could that be done for these fungi?

Steglich: I think this is only possible with wood-rotting fungi which you can grow on logs, like *Lentinus edodes* (shiitake) in Japan. The Japanese try very hard to grow *Tricholoma matsutake* (matsutake), which can cost $40 per kg. Some of the fungi, such as *Lactarius*, do not even grow in mycelial cultures; they need some special compounds from the trees with which they live in association.

Yamada: We are culturing *T. matsutake* mushrooms; the mycelia grow very well but these cultures do not produce cinnamate methyl ester or matsutakeol (1-octen-3-ol) which is an essential component of the fragrance.

Steglich: This is similar to our experience. The fruit bodies have their own metabolism. They are very rich in enzymes—anybody who is interested in enzymes should make use of this. They produce in 2–3 days enormous amounts of secondary metabolites, so it would be fascinating to do some enzyme work on fruit bodies.

Ley: How does the chemical content vary under the different growing conditions?

Steglich: We have compared mushrooms which we collected in Australia with the same species found in Germany and there was very little change in the relative composition of the compound mixtures or in the absolute values. The chemical content depends on the age of the samples; young fresh specimens are much better to work with than the aged fruit bodies, which sometimes have a tendency to form artifacts. There is also the problem of how to conserve the fruit bodies. Sometimes they degenerate very quickly, within a few hours, and we have to throw them away, it's very disappointing.

It has been suggested that organic chemists shouldn't be allowed to collect plants because of limited botanical knowledge, but I go to several places in the world and collect mushrooms with only limited assistance from microbiologists. I always keep voucher specimens for confirmation by mycologists once I get home.

Balick: Most people realize that one of the responsibilities of a botanical garden is to maintain collections of higher plants for research as well as for the enjoyment of the public. Very few people realize that we also maintain living collections of lower plants. At the New York Botanical Gardens we have some 7000 or 8000 cultures of ascomycetes and basidiomycetes. If anyone wanted to use them for research, we would be happy to make them available.

Rickards: We should emphasize the novelty of molecular structure which is available from lower plant life, and the potential to find novel pharmacological action there. I would like to relate Professor Steglich's talk to what we have already heard today and ask the ethnobotanists and ethnopharmacologists to what extent in their experience do lower plants feature in folk medicine.

Cox: In my experience, fungi are used among indigenous peoples only for psychoactive purposes. In the South Pacific and in Malaysia, *Copelandia cyanescens* is used sometimes for psychoactive purposes. It contains a tremendous amount of psilocin. This is an exception, though, and I have often wondered why more fungi are not used by people in the Pacific basin in their medicine. They are eaten a great deal and in China they are used more in medicine.

Balick: In the Amazon, you see little use of mushrooms as psychoactive agents; that's limited to Mexico and northern Central America. I think the field of ethnomycology is just not developed in the area of the New World where I work. Some of the Indians use mushrooms as antibiotics, or for dusting on wounds to promote healing, but most of the known use is for consumption rather than medicine. But I believe it's because this subject has not been studied sufficiently.

Fowler: Professor Steglich, I was intrigued by your biosynthetic pathways and the number of dehydroxylation potentials which occur. Have you looked at the enzyme systems for those reactions and do you find a specificity?

Steglich: No, this is such an undeveloped field of natural product chemistry that we still have to work out what compounds are present in the mushrooms.

Fowler: There is an important point that we have missed so far. We have been talking about the final product of synthetic pathways for different drugs. What is also important is the enzyme potentials behind those systems for biotransformations. We should look at these fungi to see if we can use the enzyme systems to modify and create bioactive products.

Steglich: Yes, it's very important and I think it's a field for the future!

Naturally occurring cyclohexene epoxides revisited

Chulabhorn Mahidol

Chulabhorn Research Institute, Bangkok, Thailand

Abstract. Cyclohexene epoxides are a small group of naturally occurring oxygenated cyclohexanes, most of which have been found in plants of the *Uvaria* genus. Some of these molecules possess pharmacological activity and have been isolated from plants that have been used in traditional medicine in Thailand. The biogenesis of these compounds was thought to have been fully explained when all the intermediates of a proposed synthetic pathway had been isolated from a single *Uvaria* species. Further studies on naturally occurring cyclohexene epoxides identified novel compounds, whose presence could not be explained by the accepted scheme. An alternative biosynthetic pathway that was formerly considered unfavourably appears to fit currently available evidence.

1990 Bioactive compounds from plants. Wiley, Chichester (Ciba Foundation Symposium 154) p 66–77

Tropical Thailand has long enjoyed the luxury of an innumerable variety of tropical plants. In this regard, the Thais have a long tradition of folklore medicine utilizing alleged medicinal herbs and plants. However, most of these 'claimed' curative properties have been neither scientifically proven nor properly investigated and it is not uncommon to find a herb cited as a cure for 'all' ailments. Research is urgently needed to explore the potential of these natural products before opportunities become forever lost through advancing deforestation.

From the many natural product research projects conducted in Thailand I would like to select one which should stimulate interesting discussion with regard to the structure, synthesis, biosynthesis and biological activity of the isolated compounds, namely the study of the 'cyclohexene epoxides' and related substances.

During 1968–1970 three cyclohexene epoxides were described, viz: crotepoxide **1** (Kupchan et al 1968), senepoxide **2** (Hollands et al 1968), and pipoxide **3** (Singh et al 1970), respectively isolated from *Croton macrostachys*, *Uvaria catocarpa* and *Piper hookeri*. The significant inhibitory activity of crotepoxide against Lewis lung carcinoma in mice (Kupchan et al 1969) attracted a great deal of attention, but it was not until ten years later that isolation of another compound of this class (from *U. purpurea*) was reported. Collaborative Thai-American

66

work proved that the latter compound was identical to the known pipoxide and further found the reported structure to be incorrect on the basis of NMR. Consequently, the structure of pipoxide was revised from **3** to **4** (note differences in both the stereochemistry of the epoxide ring and the position of the benzoate group) and confirmed by synthesis and X-ray crystallography (Holbert et al 1979).

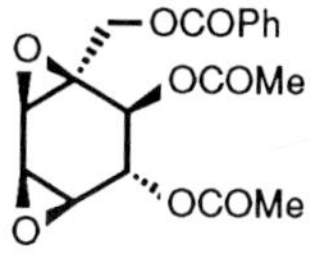

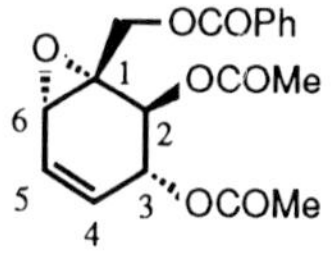

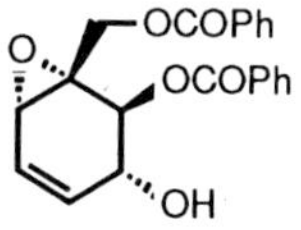

Meanwhile (prior to the revision of the pipoxide structure), an attractive biogenetic route to crotepoxide **1** from isochorismic acid **5** had been put forward by Ganem, shown in Scheme 1 (Ganem & Holbert 1977). An unfortunate drawback of the scheme was that the proposed arene oxide intermediate **7** seemed

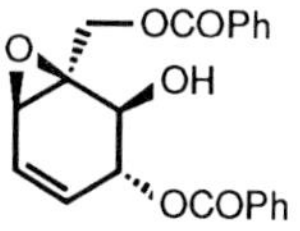

Scheme 1

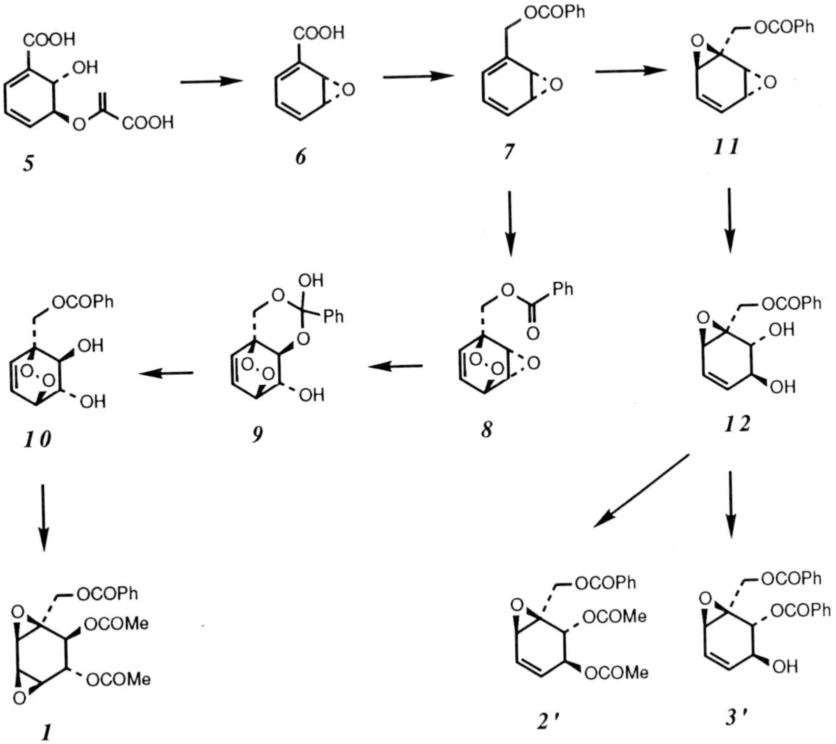

to lead to the production of **2′** and **3′** that would have opposite configurations to compounds **2** and **3** which are found in nature. Hence not much attention was paid to the pathway, which, in any case, was for a very small group of natural products with only three known members at the time.

The early 1980s saw a revival of intense interest in the chemistry of naturally occurring cyclohexene epoxides, after the isolation of a number of biologically active flavanones and dihydrochalcones containing *o*-hydroxybenzyl substituents, e.g. **17**, **18**, from species of *Uvaria*, a genus rich in cyclohexene epoxides (Hufford & Lasswell 1976, Cole et al 1976). Then, from *U. zeylanica*, the group of Cole & Bates isolated zeylena **15**, whose structure indicated intramolecular Diels-Alder addition within a diene precursor, **14a** (Jolad et al 1981). This led Cole & Bates to propose a beautifully simple biogenetic pathway from arene oxide **7′**, shown in Scheme 2. Benzyl benzoate would be epoxidized enzymically to

Scheme 2

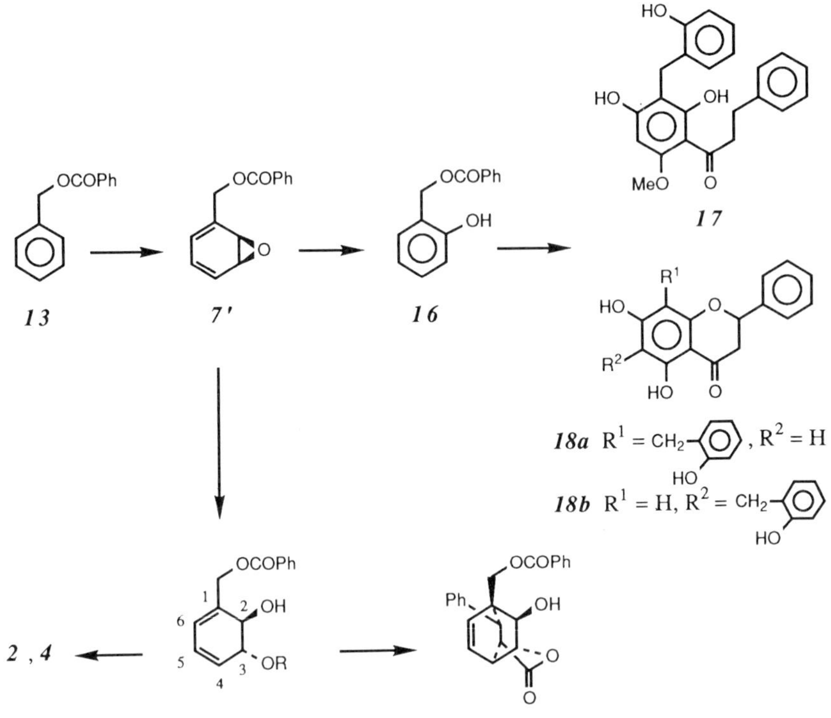

14a R = COCH=CHPh

14b R = COMe

14c R = COPh

form **7'**, which could then undergo ring-opening by various nucleophiles to give **14**. In the case of **14a**, intramolecular cycloaddition could lead to **15**.

The above pathway rationalizes not only the biogenesis of zeylena, but also the formation of senepoxide **2** and pipoxide **4**. The different stereochemistry of the epoxide rings can be explained in terms of equally feasible non-stereospecific α- and β-epoxidations at the 1,6 position of intermediate **14**. Furthermore, the prospect of arene oxide–phenol rearrangement is appealing, because it would provide a rich pool of *o*-hydroxybenzyl groups for plants in the *Uvaria* genus.

Further investigation of *U. purpurea* and *U. ferruginea* which are indigenous to the south and northeast of Thailand, respectively, yielded two new members of the cyclohexene epoxide family, i.e. β-senepoxide **19** and tingtanoxide **20** (Kodpinid et al 1983). Again, both compounds possess *2S,3R* absolute configurations similar to those of the already known members **1**, **2** and **4**. Interestingly, β-senepoxide **19** was co-isolated with its forerunner, the α-isomer **2**, which strongly supports the proposal that epoxidation occurs in the final step on either face of an intermediate diene **14** (Scheme 2).

Even more exciting was the isolation of the 'missing link' in the biogenetic pathway of Scheme 2, the intermediate dienes **14c**, **21** and **22** (Kodpinid et al 1983), together with nine benzyl benzoates, e.g. **13** and **16**, and two *o*-hydroxybenzylflavanones, **18a** and **18b** (Kodpinid et al 1984, Kodpinid et al 1985). Thus, with the exception of arene oxide **7'**, all the metabolites in Scheme 2 were found in *U. ferruginea*. Hence, despite the lack of proper feeding experiments, Cole & Bates' proposal seemed to be on firm ground.

19   *20*   *21* R = COMe

                                                                        22 R = COPh

Biomimetic synthesis of cyclohexene epoxides

In contrast to the numerous studies on the synthesis of cyclohexene epoxides in the literature, we took advantage of a biomimetic approach to their synthesis. With ready supplies of dienes **14c**, **21** and **22** from *U. purpurea* and *U. ferruginea*, we were able to carry out simple undergraduate chemistry manipulations on these advanced intermediates and convert them into the various natural cyclohexene epoxides as outlined in Scheme 3.

Scheme 3

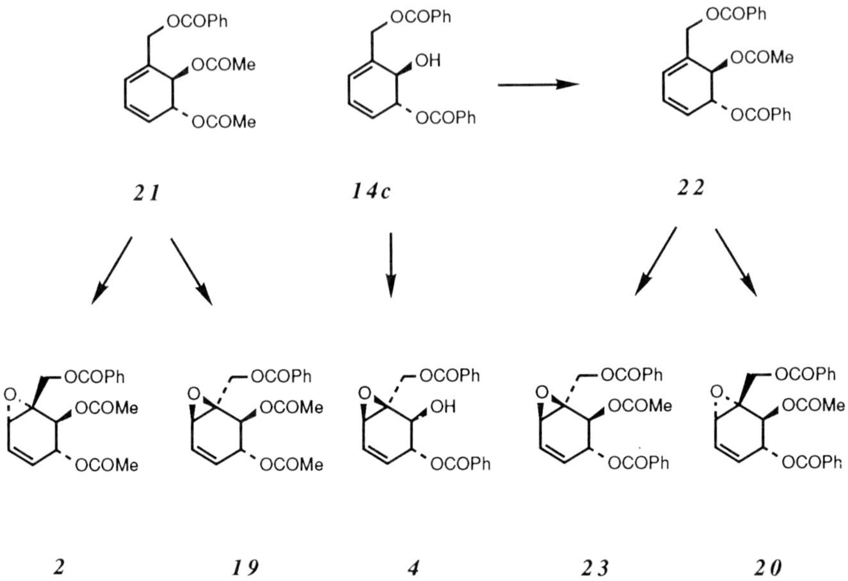

We showed convincingly that, without the directive influence of a free -OH group, epoxidation takes place on both faces of the intermediate diene. In this manner, epoxidation of dienes **21** and **22** (the latter also obtainable from acetylation of **14c**) by metachloroperoxybenzoic acid (MCPBA) gave mixtures of α- and β-epoxides in the forms of α- and β-senepoxides (**2** and **19**) and tingtanoxide and pipoxide acetate (**20** and **23**), respectively. On the other hand, the reaction of MCPBA with **14c** gave only the β-oxide, pipoxide **4**.

Further studies of the chemistry of dienes **14c**, **21** and **22** provided more information on their possible roles in biosynthesis. Exhausive hydrolysis of **14c**, **21** and **22** gave the common triol **24**, whose rate of acylation follows the order: primary 1′-OH > secondary 3-OH > secondary 2-OH. Epoxidation of **24** is both regio- and stereospecific, occurring at the 1,6-π bond and on the same face as the 2-OH group (cf. **14c**→**4**). As formerly observed with dienes **21** and **22**, protection of the 2-OH group destroys the stereospecificity to give both α- and β-epoxides. On the basis of these observations, we were able to synthesize various naturally occurring cyclohexene derivatives from triol **24**, as shown in Scheme 4.

Scheme 4

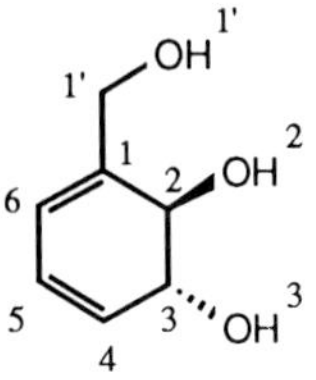

Acylation   1'-OH > 3-OH > 2-OH

Epoxidation   i) 1,6 double bond > 4,5 double bond
                 ii) β-epoxide (*cis*-to free 2-OH)
                 iii) both α- and β-epoxides
                      when 2-OH is blocked

24

i) PhCOCl (1eq) / NEt₃
————————————————————→   **15** (49%)
ii) PhCH=CHCOCl (1eq) / NEt₃

i) PhCOCl (2eq) / NEt₃
————————————————————→   **4** (56%)
ii) MCPBA (1eq)

24

i) PhCOCl (2eq) / NEt₃
————————————————————→   **23** (20%) + **20** (17%)
ii) MeCOCl (1eq) / NEt₃
iii) MCPBA (1eq)

i) PhCOCl (1eq) / NEt₃
————————————————————→   **19** (17%) + **2** (15%)
ii) MeCOCl (2eq) / NEt₃
iii) MCPBA (1eq)

Ganem's biogenesis revisited?

At this stage it looked as though the case was closed on cyclohexene epoxides and by 1986 it was generally considered that isolation of any new members of this class would simply be a variation on the theme (Thebtaranonth & Thebtaranonth 1986). How unexpected then, when the following year saw the isolation of the enantiomer, **4'** of pipoxide **4** from *U. pandensis* (Nkunya et al 1987), the first cyclohexene epoxide with *2R,3S* configuration to be found in Nature. Even more dramatic was the report by Thai chemists on the investigation of *Kaempferia* rhizomes which yielded crotepoxide **1**, boesenboxide **25**, (−)-pipoxide **4'** and its metabolites **26** and **27** (Pancharoen et al 1989).

1 R = COMe

25 R = COPh

4'

26 R = H

27 R = COPh

What is the implication of the co-isolation of *2R,3S*-(−)-pipoxide **4'** and its metabolites **26** and **27** with *2S, 3R*-crotepoxide **1** and boesenboxide **25**? Does *K. angustifolia* synthesize these *cis*-diepoxides and monoepoxide with different configurations from the same precursor, i.e. arene oxide? If not, what reason would there be (assuming that arene oxide is the true precursor) for the plant to make one arene oxide specifically for the production of **1** and **25** and another for **4'**, **26** and **27**? Of course, if we were to invoke non-regiospecific opening of the epoxide ring by nucleophiles at either C3 or (highly unlikely) C2, then any single arene oxide could give rise to both *2R, 3S*- and *2S, 3R*-configurations (Scheme 5), but how probable is such random epoxide ring opening?

Scheme 5

i) C_2 attack by OH

ii) acylation

7

7'

C_3, β-attack

C_3, α-attack

2R , 3S

2S , 3R

While we were contemplating these questions, the final twist in our line of thinking came when we re-inspected Ganem's biogenetic route. According to Scheme 1, if *Kaempferia* produced arene oxide **7**, whether from isochorismic acid or not, it would lead to the formation of *cis*-diepoxides, e.g. **1** and **25** and monoepoxides, e.g. **4'** and metabolites **26**, **27**, with opposite configurations at C2 and C3. Ironically, when Ganem made the proposal it was considered a weak point that the route should lead to *cis*-diepoxide, e.g. **1** and monoepoxides, e.g. **2'** and **3'**, with opposite configurations at C2 and C3; so much so that he suggested the production of antipodal *2R,3R*-isochorismate or the isomerization of arene oxide **7** to **7'** via an oxepin intermediate. How happy Professor Ganem would have been if the constituents of *K. angustifolia* had been investigated before he proposed the biosynthetic route.

After all that has been said and done, it still remains for feeding experiments to clarify the true pathway of production of cyclohexene epoxides in plants.

References

Cole JR, Torrance SJ, Wiedhopf RM, Arora SK, Bates RB 1976 Uvaretin, a new antitumor agent from *Uvaria acuminata* (Annonaceae). J Org Chem 41:1852–1855

Ganem B, Holbert GW 1977 Arene oxides in biosynthesis. On the origin of crotepoxide, senepoxide, and pipoxide. Bioorg Chem 6:393–396

Holbert GW, Ganem B, Engen DV et al 1979 Shikimate-derived metabolites revised structure and total synthesis of pipoxide. Tetrahedron Lett 20:715–718

Hollands R, Becher D, Gaudemer A, Polonsky J 1968 Etude des constituants des fruits *D'Uvaria catocarpa* (Annonacee). Structure du senepoxyde et du seneol. Tetrahedron 24:1633–1650

Hufford CD, Lasswell WL Jr 1976 Uvaretin and isouvaretin. Two novel cytotoxic C-benzylflavanones from *Uvaria chamae* L. J Org Chem 41:1297–1298

Jolad SD, Hoffmann JJ, Schram KH, Cole JR, Tempesta MS, Bates RB 1981 Structures of zeylenol and zeylena, constituents of *Uvaria zeylanica* (Annonaceae). J Org Chem 46:4267–4272

Kodpinid M, Sadavongvivad C, Thebtaranonth C, Thebtaranonth Y 1983 Structures of β-senepoxide, tingtanoxide, and their diene precursors. Constituents of *Uvaria ferruginea*. Tetrahedron Lett 24:2019–2022

Kodpinid M, Sadavongvivad C, Thebtaranonth C, Thebtaranonth Y 1984 Benzyl benzoates from the root of *Uvaria purpurea*. Phytochemistry (Oxf) 23:199–200

Kodpinid M, Thebtaranonth C, Thebtaranonth Y 1985 Benzyl benzoates and *o*-hydroxybenzyl flavanones from *Uvaria ferruginea*. Phytochemistry (Oxf) 24:3071–3072

Kupchan SM, Hemingway RJ, Coggon P, McPhail AT, Sim GA 1968 Crotepoxide, a novel cyclohexane diepoxide tumor inhibitor from *Croton macrostachys*. J Am Chem Soc 90:2982–2983

Kupchan SM, Hemingway RJ, Smith RM 1969 Tumor inhibitors. Crotepoxide, a novel cyclohexane diepoxide tumor inhibitor from *Croton macrostachys*. J Org Chem 34:3898–3902

Nkunya MHH, Weenen H, Koyl NJ, Thijs L, Zwanenburg B 1987 Cyclohexene epoxides, (+)-pandoxide, (+)-β-senepoxide and (−)-pipoxide, from *Uvaria pandensis*. Phytochemistry 26:2563–2565

Pancharoen O, Tuntiwachwuttikul P, Taylor WC 1989 Cyclohexane oxide derivatives from *Kaempferia angustifolia* and *Kaempferia* species. Phytochemistry 28:1143–1148

Singh J, Dhar KL, Atal CK 1970 Studies on the genus piper-X. Structure of pipoxide. A new cyclohexene epoxide from *P. hookeri* Linn. Tetrahedron 26:4403–4406

Thebtaranonth C, Thebtaranonth Y 1986 Naturally occurring cyclohexene oxides. Acc Chem Res 19:84–90

DISCUSSION

Battersby: Your Royal Highness, thank you very much for that lovely chemistry, which illustrated selectivity, stereospecificity and regiospecificity. You mentioned the biosynthetic aspects; my feeling was that your results on the isolation of the missing diene and so many of the related compounds support rather strongly the Cole & Bates hypothesis. Do you in your Institute plan to do precursor feeding experiments?

Mahidol: It is planned to do that in collaboration with another laboratory.

Mander: It seems that with many of these compounds there must be a very strong potential for them to aromatize and form phenols. Is this a difficulty in manipulating these compounds?

Mahidol: It is rather difficult to manipulate some of these compounds because they are very sensitive, presumably because of the highly functionalized epoxide ring.

Ley: May I ask Your Royal Highness about the oxidation reaction? Is there any evidence for the arene oxide intermediate or could there be a direct production of the cyclohexadiene diol by a dioxidase mechanism?

Mahidol: At present we have no evidence to support either of those mechanisms.

Ley: If one uses microbial dioxidases, one would normally get *cis* stereochemistry by oxidation of the aromatic ring. Maybe that's the reason that you have an arene oxide in your studies. I don't know whether anybody has evidence for an arene oxide in these mechanisms.

Mahidol: That is a very interesting point; in my laboratory we haven't tried microbial oxidation.

Battersby: It would be extremely interesting to know the origin of the oxygen atoms, but experimentally that would perhaps not be so easy with a higher plant. This information would discriminate cleanly between the various possibilities.

Barz: It is very likely that the oxygen atoms of the epoxide rings are derived by catalysis with a monooxygenase. Monooxygenases from plants are relatively easy to work with; they are microsomal enzymes and microsomes can be prepared quite easily from leaf, stem or root tissue. Furthermore, I am certain that the *trans*-dihydrodiol is not generated from a dioxygenase, because all evidence shows that if you have a dioxygenase you should have a *cis*-configurated and not a *trans*-configurated compound.

Fowler: Wolfgang, do you mean a cytochrome P_{450}-mediated system?

Barz: Yes, plant monooxygenases belong to the cytochrome P_{450} monooxygenase systems.

Battersby: The point I was making is that in one case a cyclic dioxygen system is present, which presumably comes from the two atoms of one molecule of oxygen.

Barz: There are several analogies in nature where you have the endoperoxide, for instance several sesquiterpene lactones where an oxygen molecule is introduced.

Battersby: Hence the point about discriminating by using isotopic labelling of the oxygen.

Steglich: These plants seem to have developed an effective pathway to benzene oxides, which are valence tautomers of oxepins. Is there any indication of the presence of oxepin derivatives in plants? Compounds of this type are known as fungal metabolites.

Mahidol: They may be present, but I have never worked with lower plants.

Ruchirawat: The oxepin intermediate was proposed by Professor Ganem for the arene oxide isomerization from one stereochemistry to the other.

Barz: Your Royal Highness, you had another very interesting example with regard to the biosynthesis. You showed that the 5,7-dihydroxyflavanone substituted in position 8 with an *o*-hydroxybenzyl moiety. From an enzymological point of view, I wonder whether this substituent in position 8 is introduced before the A ring of the flavanone is closed or if this substituent is introduced after formation of this ring. Have you found in your plant material the 5,7-dihydroxyflavanone without the substituent in position 8?

Mahidol: No, we haven't found that.

Barz: This is good evidence that the substituent in position 8 is introduced before or during the formation of the flavanone skeleton.

Rickards: Your Royal Highness, there are some extremely interesting biosynthetic questions involved here. As our chairman has pointed out, oxygen labelling would be most informative. It would be much easier to do such labelling if these compounds were produced in tissue culture rather than in whole plants. Has any work been done on tissue culture of the plants you have described?

Mahidol: Not in Thailand, because tissue culture is a new technique in our country and we lack experience in this field.

Battersby: I would strongly encourage someone to try a simple incorporation experiment, using ^{14}C-labelled benzyl alcohol and possibly some derivatives of it.

Mahidol: We will certainly try it!

Barz: If I understood the biosynthetic alternative correctly, your compounds are either formed from chorismic acid directly, or through a pathway which requires the action of phenylalanine ammonia lyase (PAL). A simple approach to distinguish whether a compound is derived from the shikimic acid pathway or through PAL is to use inhibitors of PAL. There are a number of very potent competitive inhibitors known. By inhibiting the deamination of phenylalanine and all subsequent steps, you could see whether chorismic acid is the direct precursor.

Cambie: Your Royal Highness, epoxides are often associated with cytotoxicity. Is there any recorded bioactivity of these compounds?

Mahidol: I mentioned earlier that crotepoxide was found to have cytotoxic activity; this was reported by the late Professor Kupchan. Later, I found boesenboxide, an analogue of crotepoxide. I have tested that compound and it has no cytotoxic activity whatsoever.

Kuć: Your Royal Highness, among the defence compounds in plants there are also examples of epoxides. There is a strong possibility that these are produced not only in response to infection but also as stress compounds in response to various physiological stresses.

Bowers: Your Royal Highness, is there any evidence that these epoxides might be generated by damage to the plant? Could they be produced in response to trauma by microorganisms or feeding by insects?

Mahidol: Personally, I don't think so, because when we collect the plants we make sure that they are healthy and not infected by any microorganisms or fungi, but we cannot be 100% sure.

Bowers: Sometimes in plants there are stored precursors, separated from activating enzymes, and the act of macerating, drying the plant and extracting it can bring the precursors and the enzymes together which might produce epoxides that would be toxic.

Mahidol: We tried fresh extraction and dry extraction of *K. angustifolia* to compare the chemical constituents and they are the same. One difficulty which arises, and which will become more common in Thailand, is that it is difficult to find this plant nowadays because of deforestation. I tried to grow the plant in my back yard to continue the research but it is quite difficult to grow. It is also very difficult to identify the plant, I had to wait for nearly two years for that plant to flower so that it could be firmly identified as *K. angustifolia*. I think plant tissue culture might help a great deal in this area.

Barz: I would not be too optimistic about the use of plant tissue cultures in this particular case. These compounds are highly lipophilic and one would predict that they are not localized in the central vacuole, as most hydrophilic compounds are. I would predict that the leaves have special organs—glands, oily hairs or such structures—where these lipophilic compounds are stored. The general experience is that lipophilic compounds are not easily formed in plant cell suspension cultures.

Fowler: I agree. We have screened a large number of cell cultures for monooxygenases and cytochrome P_{450} systems. We rarely find traces of the sort of compounds that you are looking at, for the following reason. You need a degree of cell and tissue differentiation and compartmentalization, as Bill Bowers suggested, to see production of these compounds. In most cell cultures you won't find that. So for once I am negative about the cell culture route— it's worth trying but I add a note of caution.

Yamada: Your Royal Highness, we have been working on the epoxidation of *l*-hyoscyamine to scopolamine. In cultured cells we couldn't find any activity of the 2-oxoglutarate-dependent dioxygenase that catalyses this epoxidation.

In root cultures, however, we found very high dioxygenase activity. In your case, the leaves accumulate tropane alkaloids, for example *l*-hyoscyamine and scopolamine, but they are generated in the roots. We confirmed this by monoclonal antibody immunological assay of the different tissues, i.e. roots, shoots and leaves. Therefore, I would encourage you, if you do not find your compounds in tissue culture, to look in root cultures.

Rickards: Your Royal Highness, I understand that the triol is a synthetic compound and has not yet been observed as a natural product, is that correct?

Mahidol: Yes, that triol is produced synthetically from the diene.

Rickards: Have you looked for that triol in the natural material using the synthetic material as a reference? The triol could be a very interesting compound; it will be less lipophilic than the esters that these people are worried about and might then be produced in tissue cultures.

Mahidol: We have not, but we will in the future.

General discussion I

McDonald: It is quite common that a natural product which is isolated and is of great biological interest may not be sufficiently robust for use: agrochemicals in particular have to have a certain potential half-life on plants. So there is a need for the modification of natural products into synthetic analogues which will give the desired effect. A very good example of that is Dr Michael Eliot's discovery of the photostable synthetic pyrethroids.

A more recent example from our laboratories came originally from a paper by Professor Steglich. He described two natural products called oudemansin and strobilurin, both with some antifungal activity. Another compound with similar activity is myxothiazole from a myxobacterium. The simplest compound of the three was strobilurin A **1**.

We used that as a starting point to look for useful antifungal activity. John Clough in our laboratory synthesized a sample of the compound. He found that it did have antifungal activity *in vitro* but it was very unstable and had no useful activity when put onto plants in normal conditions. John Clough and Mike Bushell had the idea of fusing onto this structure an additional benzene ring to make a more stable stilbene structure **2**. Professor Steglich and scientists at BASF had a very similar idea, unknown to us. Quite independently, the ICI group and the BASF group synthesized the stilbene compound and subsequently patented those compounds as antifungal agents. Then in a further step a series of antifungal diphenyl ethers was discovered **3**.

This example shows how one can start with a natural product with very interesting biological activity and from that, by introducing chemical structures that confer stability, develop a commercial product. This is the early stage of that story.

Barz: I would like to ask about the stilbene. A number of plants produce stilbenes as phytoalexins, but these are not really stable compounds, they dimerize or trimerize. For instance, the grape phytoalexins, viniferius, are derived from stilbenes. So if you have the stilbene as an antifungal compound, is it the monomer, dimer or trimer?

McDonald: The stilbene that I described has better photostability than the diene that is the natural product. That is not to say that it has sufficient stability to be used commercially. I am quite sure that if you shine light upon it, it will undergo photochemical reactions. We believed at the time that the biological activity we were seeing under low light intensity in the glasshouse was the consequence of the parent stilbene structure and not of dimers.

1

2

3

Steglich: But as you mentioned there can be a lot of variations—the stilbene is quite an active variety. It is interesting that the ICI group and our group together with BASF had the same basic idea to incorporate this benzene ring, which was the major breakthrough in this field.

The story continues. We have isolated recently a new type of strobilurin that incorporates a spiroacetal functionality in its structure. These compounds are very strong inhibitors of growth of some tumour cell cultures and also prohibit virus propagation in human cell cultures. They are all very specific respiratory inhibitors which interfere with the mitochondrial electron transport chain in the bc_1 complex. Interestingly, strobilurins and oudemansins appear not to be very toxic to animals.

Bellus: You mentioned strobilurins as new antifungal natural products. Do ICI and BASF plan to develop some of the photostable strobilurin derivatives containing an additional aromatic ring for use in agriculture as fungicides?

McDonald: We hope to do so but that isn't the case at the moment. The attractive feature is that in the group of molecules there appears to be a toxophore which already has some selectivity against fungi.

Synthesis of antifeedants for insects: novel behaviour-modifying chemicals from plants

Steven V. Ley

Department of Chemistry, Imperial College of Science, Technology and Medicine, London SW7 2AY, UK

Abstract. The need to protect our food supply from predatory insect attack using more ecologically acceptable methods has led to a rapidly growing interest in behaviour-modifying chemicals from natural sources. Compounds which deter insects from feeding (antifeedants) are attracting special attention owing to their potential use in integrated pest control management systems. We have been studying the synthesis of plant-derived compounds that display antifeedant properties. The aim of the work is to understand more precisely the feeding mechanisms of insects at a molecular level so as ultimately to be able to design simpler compounds capable of mimicking the activity of the natural products. Our synthetic studies on the sesquiterpenoid antifeedants, polygodial and warburganal, and on the diterpenoid clerodane, ajugarin I, have shown the promise of this approach. Current effort is directed towards structure–activity studies and synthesis of the extraordinarily potent antifeedant and growth-disrupting agent azadirachtin, isolated from the Neem tree, *Azadirachta indica* (A. Juss). This work has led to the correct structure assignment for azadirachtin and afforded many compounds for biological evaluation. It is of special significance that incorporation of the hydroxydihydrofuran portion of azadirachtin in a simple model system formed a compound with antifeedant activity comparable to that of the natural product.

1990 Bioactive compounds from plants. Wiley, Chichester (Ciba Foundation Symposium 154) p 80–98

Despite an enormous expenditure on agrochemicals worldwide there are still considerable losses of crops owing to damage caused by feeding insects and other pests. Current environmental pressures and rapidly developing resistance to conventional insecticides provide the impetus to study new, more ecologically acceptable methods of pest control. Plants have evolved highly elaborate chemical defences against attack and these have provided a rich source of biologically active compounds which may be used as novel crop-protecting agents. The development of the pyrethroids from the natural pyrethrin is a classic example of these studies. More recently, interest has been directed towards behaviour-modifying chemicals that can be used in integrated pest control management programmes. Of these, compounds which affect the feeding processes of insects are becoming increasingly important. However, these so-called antifeedants are only just beginning to meet the rigorous requirements necessary for commercial development (de Groot 1986).

We became interested in the synthesis of antifeedants several years ago. Our aims were to devise routes to the natural products such that we could probe the necessary functional groups in the molecules required for biological activity. Furthermore, these studies would contribute to the fundamental understanding of feeding mechanisms and taste receptors at a molecular level. By careful entomological evaluation we hoped that the research would ultimately provide a basis for the design of structurally novel and simpler compounds which would be capable of mimicking the natural products. We also hoped to incorporate additional desirable features into these molecules, for example effects on growth regulation or inhibition of oviposition.

During the early synthetic studies, the work focused on developing highly efficient routes to the drimane antifeedants, polygodial and warburganal (Ley et al 1983). Additionally, several analogue compounds were prepared and their antifeedant properties assessed. One of these, formed in only three steps from readily available starting materials displayed a level of activity equivalent to polygodial itself against four insect species (Blaney et al 1987).

Polygodial   Warburganal   Analogue Compound

Further synthetic studies on the structurally more complex clerodane diterpene antifeedants culminated in the total synthesis of ajugarin I (Ley et al 1986), the active component isolated from the *Ajuga remota* plant. During this work, a detailed structure–activity study was undertaken, including preparation of simple analogues and chemically modified natural products, and comparisons with related natural materials. These studies highlighted the essential structural requirements of the epoxide and the two ester groups in the decalin portion of the molecule in precise stereochemical arrangements. They also showed the additional importance of polar side chains attached to the decalin, which could impart a synergistic improvement to the activity. However, structural modifications in the side chain led to different antifeedant behaviours, depending on the insect species (Ley et al 1986, Blaney et al 1988).

Ajugarin 1

In the clerodane area, recent work with our colleagues at the Jodrell Laboratory at Kew has led to the isolation of five new compounds from the plant genus *Scutellaria*, the structures of which are shown below (Anderson et al 1989, Cole et al 1990). While we have not yet assessed all these for antifeedant activity, it was observed that jodrelin B was 5 to 10 times more active in our screens that any previously isolated clerodane. This result could have important implications in the design of novel antifeedants.

R=Ac Jodrellin A
R=CO.^{i}Pr Jodrellin B

X-Y= CH=CH Jodrellin C
X-Y=CH$_2$-CH$_2$14,15-Dihydrojodrellin C

Galericulin

Without question, however, the most potent antifeedant yet discovered is the molecule azadirachtin. This compound isolated from the Neem tree, *Azadirachta indica* (A. Juss) (Morgan & Butterworth 1968) has been the centre of considerable research effort (Jacobson 1989). After the correct structural assignment of azadirachtin (Ley et al 1987a, Kraus et al 1987, Nakanishi et al 1987), the stage is set for structure–activity studies and synthesis.

Azadirachtin

We have already shown that azadirachtin will undergo a number of useful chemical transformations. While some of these will be discussed here, their biological evaluation as antifeedants and growth disrupting agents is presented elsewhere (Blaney et al 1990, Simmonds et al 1990). The first reaction we studied involved stepwise hydrogenation of the two double bonds in azadirachtin. This process was important because the molecules derived were considerably more stable than the parent compound yet retained its biological profile. Other useful reactions involved removal of the C1 tigloyl ester grouping by oxidation and hydrolysis (Ley et al 1989b). This provided useful compounds for further analogue preparation and structure determination (Ley et al 1987a).

Since we knew that the C22–C23 enol double bond in azadirachtin was extremely labile, we investigated reactions at this site. In particular, we found that addition reactions with alcohols in the presence of bromine or by the use of acetic acid proceeded smoothly (Ley et al 1988b). Once again the products of these reactions proved to be of great value in the structure–activity studies. In general terms it was found that groups at the C22 and C23 positions lowered the biological activity and that activity dropped dramatically with increasing steric bulk of the introduced substituents. These addition products were useful for interconversion to other natural products not readily available by extraction from the tree. Also, it was shown that azadirachtin could be recovered quantitatively by pyrolysis *in vacuo* of the acetic acid adduct. This observation is likely to be important during synthetic studies because recovery of the sensitive enol double bond must be achieved at a late stage in the synthesis.

$170°$ 1.3×10^{-3} mmHg

5 min   99%

Azadirachtin

The dihydroazadirachtin was also shown to undergo rearrangement reactions in the presence of acid catalysts such as amberlyst 15 ion exchange resin and 4 Å molecular sieves (Scheme 1). This rearrangement involves ring opening of the epoxide at the C13 position by the hindered C7 hydroxyl group (Ley et al 1988a). Similar azadirachtin rearranged products have also been observed as natural products (Kraus et al 1987). Noticeable in this rearrangement is that, once the strong hydrogen bond between the C11 OH and the epoxide ring is broken, one obtains an equilibrating mixture of C11 hemiacetals.

Scheme 1

Amberlyst A15
RT   62%
4Å Selves

1 ⇌ 3

The other important component of the Neem tree extract which shows comparable biological activity to azadirachtin is the 3-tigloyl-11-deoxy azadirachtin. This compound undergoes similar chemistry to azadirachtin as reported above and has been discussed in some detail in other publications (Ley et al 1989b). All these skeletal and structural modifications provide materials for biological evaluation. Additionally, much information has been gained regarding handling and relay studies that will be vital during our total synthesis programme of these important compounds. For this work we are setting ourselves a formidable challenge, since azadirachtin contains no less than 16 stereogenic centres and a plethora of chemically sensitive oxygen functionalities. Nevertheless, this synthesis work has begun and we have made significant progress and achieved several goals. Our general plan is to effect a convergent synthesis, bringing together two important fragments, a decalin unit and a protected hydroxydihydrofuran acetal, to form the key C8–C14 bond. While many opportunities present themselves for this coupling, that outlined below contains interesting chemistry and substituents which we believe may be useful in effecting the final transformations.

To this end, we have devised a strategy which affords appropriately substituted decalins and have reported these preparations in full (Ley et al 1989a). Construction of the necessary optically pure tricyclic hydroxyhydrofuran

Azadirachtin   Synthesis

fragment is also well advanced. For this work we have used the readily available (−)-bicyclo[3.2.0]hept-2-en-6-one as a starting point and converted this via a process similar to that used in a model system (Ley & Anderson 1990) to an advanced pivotal intermediate. This compound will be of use in the total synthesis studies and in derivatization to novel antifeedant materials.

In accord with the previous observations, that fragments of the natural products can express biological activity and that the structure–activity studies suggested the importance of the tricyclic hydroxyfuran unit, we have prepared model compounds **1** and **2**.

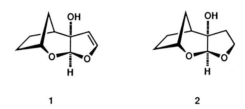

1   2

Interestingly, these simple compounds do show antifeedant activity and compound **1** is very active and comparable with the natural product against some insect species, especially the Egyptian cotton leaf worm, *Spodoptera littoralis* (Ley et al 1987b, Blaney et al 1990, Simmonds 1990). These results are very encouraging for the future design of novel antifeedants and may well provide a new approach to insect pest control as part of an integrated management programme.

References

Anderson JC, Ley SV 1989 The structure of two new clerodane diterpenoid potent insect antifeedants from *Scutellaria woronowii* (Juss): jodrellin A and B. Tetrahedron Lett 30:4737–4740

Blaney WM, Simmonds MSJ, Ley SV, Katz RB 1987 An electrophysiological and behavioural study of insect antifeedant properties of natural and synthetic drimane related compounds. Physiol Entomol 12:281–291

Blaney WM, Simmonds MSJ, Ley SV, Jones PS 1988 Insect antifeedants: a behavioural and electrophysiological investigation of natural and synthetically derived clerodane diterpenoids. Entomol Exp Appl 46:267–274

Blaney WM, Simmonds MSJ, Ley SV, Anderson JC, Toogood PL 1990 Antifeedant effects of azadirachtin and structurally related compounds on lepidopterous larvae. Entomologia 55:149–160

Cole MD, Anderson JC, Blaney WM et al 1990 Novel neo-clerodane insect antifeedants from *Scutellaria galericulate*. Phytochemistry 29:1793–1796

de Groot Ae 1986 Terpenoid antifeedants, part 1. An overview of terpenoid antifeedants of natural origin. Recl Trav Chim Pays-Bas Belg 105:513–527

Jacobson M 1989 Focus on phytochemical pesticides, vol 1. The Neem tree. CRC Press, Boca Raton, Florida

Kraus W, Bokel M, Bruhn A et al 1987 Structure determination by N.M.R. Azadirachtin and related compounds from *Azadirachtin indica* A. Juss (Meliaceae). Tetrahedron 43:2817–2830

Ley SV, Anderson JC 1990 Chemistry of insect antifeedants from *Azadirachta indica* (Part 6): synthesis of an optically pure acetal intermediate for potential use in the synthesis of azadirachtin and novel antifeedants. Tetrahedron Lett 31:431–421

Ley SV, Hollinshead DM, Howell SC, Mahon M, Ratcliffe NM, Worthington PA 1983 The Diels-Alder route to drimane related sesquiterpenes; synthesis of cinnamolide, polygodial, isodrimeninol, drimenin and warburganal. J Chem Soc Perkin Trans I p 1579–1589

Ley SV, Jones PS, Simpkin NS, Whittle AJ 1986 Total synthesis of the insect antifeedant ajugarin I and degradation studies of related clerodane diterpenoids. Tetrahedron 42:6519–6534

Ley SV, Bilton JN, Broughton HB et al 1987a An X-ray crystallographic, mass spectroscopic and N.M.R. study of the limonoid insect antifeedant azadirachtin and related derivatives. Tetrahedron 43:2805–2815

Ley SV, Santafianos D, Blaney WM, Simmonds MSJ 1987b Synthesis of a hydroxy dihydrofuran acetal related to azadirachtin: a potent insect antifeedant. Tetrahedron Lett 28:221–224

Ley SV, Bilton JN, Jones PS 1988a Chemistry of insect antifeedants from *Azadirachta indica* (Part 1): conversion from the azadirachtin to the azadirachtinin skeletons. Tetrahedron Lett 29:1849–1852

Ley SV, Anderson JC, Blaney WM et al 1988b Chemistry of insect antifeedants from *Azadirachta indica* (Part 3): conversion from the azadirachtin to the azadirachtinin skeletons. Tetrahedron Lett 29:5433–5436

Ley SV, Somovilla AA, Broughton HB et al 1989a Chemistry of insect antifeedant from *Azadirachta indica* (Part 4): synthesis towards the limonoid azadirachtin; preparation of a functionalised decalin fragment. Tetrahedron 45:2143–2164

Ley SV, Anderson JC, Blaney WM et al 1989b Insect antifeedants from *Azadirachta indica* (Part 5): chemical modification and structure activity relationships of azadirachtin and some related limonoids. Tetrahedron 45:5175–5192

Morgan ED, Butterworth JH 1968 Isolation of a substance that suppresses feeding in locusts. J Chem Soc Chem Commun p 23–24

Nakanishi K, Turner CJ, Tempester MS 1987 An N.M.R. spectroscopic study of azadirachtin and its trimethyl ether. Tetrahedron 43:2798–2803

Simmonds MSJ, Blaney WM, Ley SV, Anderson JC, Toogood PL 1990 Azadirachtin: structural requirements for reducing growth and increasing mortality in lepidopterous larvae. Entomologia 55:169–181

DISCUSSION

Farnsworth: It is interesting that all this work began with an ethnobotanical observation that natives used parts of this tree as an insecticide. All these compounds seem to have epoxides or the potential to be metabolized to epoxides; have they been tested for mutagenicity?

Ley: We have provided compounds to be screened for all sorts of biological activity, including to see whether they would bind to DNA, similar to the aflatoxins and materials of that type. I am happy to say they are completely free of any such binding properties. They have been in Ames tests and a number of other toxicity screens. Indeed, they were inactive in most of the biological screens we studied. The pharmacology and toxicology of Neem extracts has also been reported (Jacobson 1988).

Fleet: Do you know how rapidly these compounds are metabolized and what the metabolic fates of the materials are? In the absence of such information, you may waste quite some time if the initial substances are metabolized into more toxic materials.

Ley: Compounds that are metabolized rapidly clearly will not stay in the insect very long and are likely to have relatively poor biological effects. Azadirachtin is not readily metabolized; it is already highly oxidized and stays in the insect in a membrane-bound form.

We are trying to find out what happens to these molecules chemically; for example, we have tried to add nucleophiles, but without success. We are also looking at binding sites in the insect. We have investigated the larval frass to see if we can find further oxidized metabolites, but have nothing to report yet.

Steglich: You showed that you could reduce a triterpenoid to a monoterpenoid and retain its antifeedant activity. Besides iridoids, are there other monoterpenoids known which have strong antifeedant activity?

Ley: Yes, but there are not many; I know of work only on *p*-cimol, terpinan-4-ol and α-terpineol (Gombos & Gasko 1977) and catharidine (Carrel & Eisner 1974).

van Montagu: Are you interested in using enzymes for some steps in this synthesis? It would be possible to clone from these plants genes for particular enzymes that could help in the synthesis.

Ley: There is great interest by synthetic chemists in using enzymes; we commonly use them in our work. In this particular synthesis programme we are hoping to introduce the C-11 hydroxyl group of azadirachtin using a microbial oxidase method. We are also looking at much less oxidized substrates to see if these materials are precursors for active compounds that can be formed by enzymic oxidation methods. There are exciting opportunities to achieve transformations that are not easy to do using conventional organic reagents.

van Montagu: We have shown that the *Bacillus thuringiensis* insecticidal protein (BT) binds to receptors with a dissociation constant of 10^{-10}, so the specificity is very high (Van Rie et al 1989). This specificity may be more difficult to achieve for chemical compounds. We fed insects for over a year with BT and no resistance evolved. There seem to be two receptors and the insect produces less of one receptor and more of another receptor. What is your idea about the future of this BT work?

Ley: We are preparing labelled compounds ourselves to study membrane binding and receptor selectivity. We are working with biologists to learn about the various receptors for these materials. We are not only interested in antifeedant effects; we are also interested in effects on growth, development and mortality. We are using whole insect NMR methods to see where labelled materials are bound. We are trying to find out how selective these compounds are for insect species. We are also working with single cell lines to determine the hormone receptors. In fact, we are collaborating with as many people as can convince us that they have interesting research programmes in this area. There is an awful lot of fundamental work to do, but, as you point out, to find the receptors and find out exactly what is going on at the molecular level is essential.

Cox: I was particularly struck by the work on the structure and biological activity of the clerodanes. The genus *Clerodendrum* is prominently used in traditional medicine in South East Asia; *C. inerma* is very frequently used in Samoa and *C. floribunda*, which is found here in Thailand, is also commonly used in folk medicine. The entire Verbenaceae, including other genera, such as *Stachytarpheta*, is prominent.

How did you choose *Clerodendrum* to work on? Secondly, what species of *Clerodendrum* did you isolate these clerodanes from?

Ley: Originally, I noticed in the literature that an incorrect structure had been reported. This was for clerodane itself, which was first reported by Sim et al (1962). It appeared to me that their X-ray crystal structure was incorrectly represented as an antipode (with the opposite chirality) to what it should have been (Rogers et al 1979). The source of this clerodane was *Clerodendrum infortunatum*, which is rather appropriate!

Although the side chains in some of those materials are quite elaborate, we believed that the epoxy diacetate functional group was important for biological activity, but nobody had shown this. We began to look for all natural products that contained this functional group arrangement. We used molecular modelling and mapping of oxygen placement within natural products to find other compounds from the X-ray crystal database, and some of these turned out to be antifeedants. In the clerodane area we began a collaboration with groups in Spain and in Sicily, who gave us a number of natural products which we put through our screens. Very few were active as antifeedants except those that had these particular structural features, of the epoxide and the primary and secondary diacetates (Simmonds et al 1989).

We wanted to study plants that we could grow locally, which is how we got into looking at the genus *Scutellaria* and the *Ajuga remota* plant. These are small crop plants that one might be able to use to produce the natural products. This would obviate the need to take down large trees to provide the natural products. Small plants also provide the opportunity of using tissue culture methods of production.

Fleet: Could I ask about the fragments you said were active when they were racemic. Does that mean that the activity that you find with the homochiral material is the same or is it different?

Ley: There is some confusion in the literature on this one. It was reported that for polygodial one of the enantiomers was phytotoxic. This was not true; the toxicity was due to a contaminant and both enantiomers turned out to be nearly equipotent against aphids (Griffith et al 1988). We don't believe this will necessarily be true in the new series of compounds; we feel that the homochiral material will be very much more potent. We have now prepared the model compound in homochiral form and this is being screened for biological activity.

Barz: What does biological inactivity mean in this sense? Does the compound no longer bind to the receptor or is it preferentially detoxified?

Ley: No, it means the insect will eat normally.

Kuć: In that context, aren't some bitter compounds actually insect attractants? For example, cucurbitacins attract the cucumber beetle. Can compounds be antifeedants for some insects and attractants for others?

Ley: I would say that you shouldn't use bitterness as the only guide to find an antifeedant compound. The azadirachtins are effective against 150 insect species and I don't know of any that are attracted to it.

Mander: Could you expand on the species specificity?

Ley: The clerodanes have a high degree of insect specificity. We get good activity against some *Spodoptera*, such as *littoralis*, but it's much more important commercially to get effects on *Heliothis*. We have found that *Heliothis armigera* is probably the most resistant of the ones that we study, although azadirachtin is good against most species.

Mander: How about larvae versus adults?

Ley: It depends on the species. Most of the adults have more interesting things on their minds, like sex, than eating. The larval stages are when most feeding occurs. In the adults many of the compounds can act as oviposition inhibitors so there is the possibility of stopping the egg-laying process. This is an added benefit of these compounds.

Potrykus: If I understood correctly, you try to spoil the appetite of the locust and other insects. How do you then prevent them going somewhere else where the plants have not been sprayed with your compounds, for example the neighbouring farm?

Ley: That's a good point. In the field there will be a combination of methods operating. There could be part of a plant, for example, that has not been covered by spraying and the insect could simply walk to that part and start eating. The best compounds that we are working with are systemic; they actually translocate within the plant and as the plant continues to grow it carries the chemical with it.

There are double spraying techniques where one would use an antifeedant early in the cycle to reduce the size of the caterpillar and then use a second insecticide spray to kill the insect. In this way, you need less of the insecticide.

In a lot of the biological screens that we do, the insect is presented with its food in a 'no choice' fashion, which is probably most like the field situation. The insect is not given the opportunity of eating anything else, to see if it can be forced to feed on the antifeedant. We have done this through 35 generations and seen no resistance build up.

Potrykus: Would it not be an advantage to kill the insects instead of just chasing them away?

Ley: Yes, if you can target the insect that you want to kill. I don't believe that any one type of therapy is the way forward, one needs a whole armoury of integrated pest control methods.

Maybe it was implied that we were driving insects away from the protected crop—that's true, but you can't drive them over great distances, so they die of starvation or dehydration. Also, many of the compounds, particularly the azadirachtins, have secondary effects and indeed are toxic. We also monitor the mortality produced by the azidiractin compounds and find that nearly all the insects die at or before the pupation stage, relatively few emerge as adults.

Bowers: I might have an answer to Dr Potrykus' question of how do we stop the insects running away. Ecdysone is the principal moulting hormone of insects and regulates their development. Recently, Wing et al (1988) published the structure of a compound, that came out of their synthetic screen, which showed some ecdysone agonist activity.

I was sent a sample of this compound and I gave it to one of my students, Felix Ortego, who was working on *Schistocerca americana*, a close cousin of the desert locust. He tested this compound, came back to me and said I want to show you something. He wouldn't tell me; he wanted to demonstrate using the insects. He was working with the last instar nymphs of this grasshopper and normally they will leap away from you. He injected the nymph, held his hand over it for 15–20 seconds then took it away. I expected the nymph to jump but it just sat there. Within a very short time it began to twist its metathoracic legs over its back into various unnatural positions, then 'clunk' one leg fell off; it twisted the other leg over its back, 'clunk' and the other leg fell off, all in about two minutes.

This has been tested on *Schistocerca gregaria* and it does the same thing, so I don't think you have to worry about *gregaria* going anywhere!

It works on the adult insects too, which are the migratory form. All of us have caught a grasshopper and had it leap away leaving a metathoracic leg in our hands. This is a process called autotomy, by which grasshoppers and crickets are able to release the leg and escape, and they can survive for a while without one leg. However, they don't do so well without two legs.

This effect on autotomy is not related to any ecdysone agonist activity; ecdysone will not produce this effect. This is obviously an effect on the nerves that control the autotomy response and it renders them essentially incapable of both jumping and flying. It will be interesting to see what synthetic chemists can make of this structure. This effect is specific for saltatory orthoptera; it is not transferable to other insects, like cockroaches.

Potrykus: I can imagine that locusts can fly hundreds of kilometres without legs and they can creep around without back legs. I think it will be difficult with caterpillars to convince them to throw away their legs; they have many and relatively short ones! Therefore, the basic question I have is, what is the specific advantage of the strategy of feed inhibition versus killing?

Ley: I think this is just one part of an integrated programme that one should adopt. If one adopts a too rigid regime of trying to kill things by one type of mechanism, resistance can build up rather rapidly. The antifeedants could operate on several different mechanisms and therefore development of resistance is likely to be slower.

The other important point is that we would like to have selectivity, we do not want to kill everything. There is no point in killing insects unnecessarily, they act as food sources for birds and other organisms, and they are important in other ways for the environment. I really don't think there is a need to design systems specifically to kill. Other strategies should be looked at, and I believe the behaviour-modifying chemicals are one approach that should be used.

There is another point about this method of control. Industrialists will say that if you have something that can cause insects to migrate from one field to another field this is not bad for business because it forces the other farmer who has accumulated all the migrating species to buy their chemicals.

Potrykus: That comment exactly reinforces my argument against feeding inhibition. Of course it is advantageous for industry to spray an entire continent to prevent the locust migrating and causing damage elsewhere. There are lots of other strategies that do not kill *all* insects. I must admit I am still not convinced by the feeding inhibition strategy.

Ley: These compounds do kill, they are not just antifeedants, they are inherently toxic materials. They can be regarded as slow-acting insecticides.

The locust plagues don't occur very often, but when they do they are very dramatic and very difficult to control with insecticides. It is important that countries which are unable to afford expensive insecticides have the opportunity to provide some sort of protection to their crops. In West Africa, millet, sorghum and maize are now being sprayed with Neem extracts and they are proving to be a very acceptable method of control at the village level. Much more acceptable than the expensive insecticidal materials.

Kuć: I would like to emphasize that it's very important that we test many approaches to controlling pests. We should not rely on what has been the classical approach of merely quick-kill of the insect. This has developed problems and if we take other approaches we may be able to minimize the difficulties that have arisen.

Bowers: In general the immatures of insects are feeding machines, the adults are breeding machines. Controlling either of those stages is useful. You want to stop the immatures from feeding and damaging agricultural crops. You want to stop the adults breeding or transmitting diseases, as is the case with mosquitos and black flies. Toxicity has been a useful approach and it seems that Nature and toxicologists have conspired to develop neurotoxins. The basic biochemistry of all organisms, however, including invertebrates and vertebrates, is very similar. Generally speaking, compounds that have rapidly acting toxicity (e.g. neurotoxins) in insects also affect vertebrates. There are many exceptions and where those exceptions can be taken advantage of they make useful research tools and sometimes good products.

The first selective methods for insect control exploited insect growth regulators, developed as modifications of insect juvenile hormones. These compounds interfere with the growth and development of the immatures when they are undergoing changes controlled by the endocrine system, especially the juvenile hormone. If you give insects juvenile hormone at a time when they shouldn't have it, they moult into intermediates, they stop feeding and they don't form normal adults. These compounds are the basis of many successful products on the market today.

Insects modulate each others behaviour by communication via pheromones. These have been useful in a number of instances, for example they can be used to diminish the ability of male adults to locate females for breeding.

I think these are acceptable methods of insect control that have no potential toxicity to vertebrates and domestic animals. Everyone involved in agrochemistry, including most companies, is looking for intrinsically non-toxic methods of managing insect populations. Toxicity is still a useful tool but it can be supplanted where research indicates non-toxic methods can achieve the same result.

Cox: Antifeedant compounds might be more environmentally benign than toxic compounds in the sense that they are less likely to enter the food chains and be biologically amplified. Do you agree with that?

Ley: I think that's probably true but I wouldn't like to make a sweeping statement about it. Many of these plants and plant materials have been part of the food chain for centuries. Parts of the Neem tree have been used as food for a long time. Until full Western-style toxicological studies have been done on all components of the extract, one should be very careful about making general statements. Very small quantities of other materials present in the sample could be what is responsible for the toxic side effects, even though the compound responsible for the effects on insects may be completely innocuous.

Bunyapraphatsara: I have a question about the Neem tree. Since you have so much experience with azadirachtin, which is unstable, do you have any suggestions about how the farmer could prepare the crude extract for application in the field without loss of activity?

Ley: This is a very good point, a lot of mistakes have been made in preparation of materials. The US Development Agency working with the W. R. Grace Company has produced a Neem extract material called Margosan O that is stable for up to two years. It is stabilized in solution by the presence of sunscreens and antioxidants (J. Walter, International Chemical Congress of Pacific Basin

Societies, Honolulu, Hawaii, December 1989). There are also tricks to the extraction procedure, if you talk to people who make these mixtures in India, they can tell you the best way of getting the most active materials from the plant. Also, it is not just azidiractin that's the active component, there are other compounds present that are active. I prefer to work with pure compounds because some of the minor products present in Neem extracts are quite toxic and we do not know what effects they will have on the environment.

Cragg: It's very interesting that these products are isolated from plants growing or generally found in developing countries. There's the potential for the development of a multimillion dollar market for an effective insecticide or antifeedant. Is there any thought given to how countries where the plants are originally found can be compensated? At the National Cancer Institute and NIH we are very conscious of this. We are going to make provisions that if some potential new drug is isolated, the country of origin will be compensated.

Ley: Companies that are investing in these programmes will be building extraction plants in India. I hope the developing countries will benefit from the natural product collection. It is very difficult, because although the natural product might have some activity, it's likely to be synthetic analogues which have the real commercial prospects. The connection with the initial source of activity is sometimes rather remote.

Bellus: What needs to be developed is a simplified Neem derivative that fulfils the following requirements: 1) it should have a specific activity against only the target organism; 2) it should be non-toxic to humans and organisms in the environment; 3) it should be used at a low rate and for a minimum number of applications (this is very important so that, for example, there will be no contamination of ground water); 4) it should be degraded to non-toxic products within an appropriate time; 5) it should have physical properties that prevent uncontrolled environmental contamination; and 6) it should not accumulate in non-target breeding systems.

To fulfil all these requirements would need an additional $80–100 million to be spent on very stringent investigation of toxicological and environmental effects.

By such an enormous effort, compounds might be created which will be safe to the inhabitants and the farmers of the countries from where the plants were taken. I would like to remind you that it took nearly sixty years of concentrated research efforts all over the world to develop photochemically stable pyrethroids from the photolabile, naturally occurring chrysanthemic esters.

At CIBA-GEIGY, and I think this applies to the agrochemical industry in general, there has been enormous progress in the quality of compounds that have entered the development pipeline since 1960 (Fig. 1). As far as acute toxicity is concerned, there has been an astonishing improvement. In 1960, carbamate and organophosphate insecticides predominated. If we look at the suitability for integrated pest management, there has also been considerable progress; the same is true for the typical application rates. Of course, we would like to introduce the most recently developed compounds to the market as soon as possible. If you think we in the agrochemical industry are being slow, do not believe that this is the fault of the industry alone. We are living in an age of inverse logic: the public wants rapid introduction of environmentally more acceptable plant protection products, but the registration authorities, which are elected by the same public, require incredibly stringent investigation of these new compounds, which takes a long time.

The interesting compounds Steve Ley is developing will certainly have to go through the whole toxicology testing, and I think this is right. But if they are proved innocent, they should be registered as soon as possible.

Kuć: Have you considered the economics of using these compounds, azadirachtin, for example? It seems to me that you would have to have a synthesis that was very direct and relatively simple to make it economical.

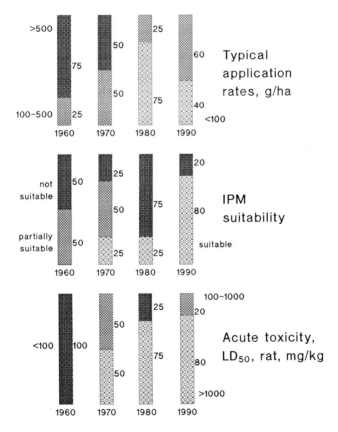

FIG. 1. *(Bellus)* CIBA-GEIGY's development compounds for insect control (numbers in bars, % of total number of compounds).

Ley: It's not really our business to consider this. The Neem extract, Margosan O, which is on the market, is considered to be quite acceptable in cost terms at $65 per gallon, especially if you exploit its selective properties, such as in a greenhouse on ornamental plants. It isn't our plan to synthesize azadirachtin in a commercially acceptable way, I don't think that would be possible. What we might eventually do is provide a novel material which can be prepared in 2–3 steps, which may be suitable for commercial development.

Rickards: Steve, you described work where you conceptually cut the molecule into two fragments, a left half and a right half in round terms, you synthesized these two halves and the left-hand segment was virtually inactive and the right-hand segment was highly active. Have you tested those two together to see if there are any cooperative or synergistic effects?

Ley: Yes, we did exactly that when we were working with the clerodane molecules. We didn't see great synergism between various fragments. When we mixed the clerodane molecules with some of the drimanoid natural products, we did see synergistic activities but they were not significant.

In the azadirachtin area, we don't see synergism between the components, which is a pity. I think it's probably because our left-hand fragment isn't good enough yet. I suspect that when we put on more lipophilic groups, the situation may be different. You are quite right; if you are stimulating only one small part of the receptor with one model compound and there are two parts to that receptor, you should use a two component mixture.

Rickards: How accurate is your assay for detecting differences between materials with similar activity?

Ley: It's pretty accurate and reproducible. We have been working in this area now for many years; our entomologist colleagues are extremely careful, they do many replications of all the data we obtain. We repeat results with different batches of insects and different insect species. One has to be very rigorous, because insects are not always the same and they can vary their behaviour from day to day. We do try to repeat our results periodically, to make sure there is consistency; it's very important to do that. We also use electrophysiological methods along with the antifeeding screens.

Our entomologists are prepared to sit down and watch these insects solidly for 12 hours. They record all their observations: whether the insects are slow to eat, whether their feeding behaviour is unusual, such as whether the insects show head rearing or wandering behaviour. We record weight changes, we look for morphological effects and lesions at the pupal stage. Anything that's unusual we try to record, because this can give a lead into novel structures. It really is a combination of all these facts that gets good results. Industrialists tend not to do this manpower-intensive type of screening, but it's our view that if anything is going to come of this, you have to be very thorough.

Bowers: These analogues of azadirachtin and some of the other terpenes, what do they taste like to humans?

Ley: I always discourage students from tasting chemicals. I don't think anybody should knowingly test these things on themselves. One student, however, did put the polygodial onto his tongue to taste it. It was extremely hot, and the effect lasted for three days. Since that time we have tasted some of the other natural products that are active as antifeedants and they do not always have an unpleasant taste.

The azadirachtins are well known to be quite bitter tasting compounds. Bitter and hot flavours are acceptable to mammalian tastes but not to insects.

Bowers: I isolated polygodials from a plant on the basis of their fungicidal activity. I tasted one and experienced some serious unpleasantness—it was very hot!

McDonald: The carbon skeleton of that very complex azadirachtin structure looks to be that of a steroid which has had ring C cleaved and four carbons chopped off the end of the steroid side chain plus, of course, lots of oxygenation. There are quite a lot of the compounds like that that are the bitter principles. You classified the azadirachtin as a limonoid. Is azadirachtin related—I am only talking of generalities—to any other bitter principle? Have any other bitter principles been tested for antifeedant or insecticidal effects?

Ley: There are groups becoming interested in the biosynthesis. You are quite right about those connections of the ring C cleavage. One of our synthetic schemes will make use of these compounds as precursors.

Certain bitter principles, for example limonin itself, are potent antifeedants (see Bentley et al 1988). It's a matter of degree; the azadirachtins are much more potent than bitter principle compounds.

Elizabetsky: Could I hear your opinion about rotenone-based formulations?

Ley: Our entomologist colleagues have studied rotenones, but we think they are rather toxic. We have some evidence that they will bind to thiols and to other amines. Because of these effects, we are less likely to develop any of them, even though they are active as antifeedants or deterrents.

References

Bentley MD, Rajab MS, Alford AR, Mendel MJ, Hassanali A 1988 Structure-activity studies of modified citrus limonoids as antifeedants for Colorado potato beetle larvae, Leptinotarsa-decemlineata. Entomol Exp & Appl 49:189–193

Carrel JE, Eisner T 1974 Cantharidin: potent feeding deterrent to insects. Science (Wash DC) 183:755–757

Gombos MA, Gasko K 1977 Extraction of natural antifeedants from the fruits of *Amorpha fruticosa* L. Acta Phytopathol Acad Sci Hung 12:349–357

Griffith DC, Lalleman JY, Ley SV et al 1988 Activity of drimane antifeedants and related-compounds against aphids, and comparative biological effects and chemical-reactivity of (−)-polygodial and (+)-polygodial. J Chem Ecol 14:1845–1855

Jacobson M 1988 Focus on phytochemical pesticides. CRC Press, Boca Raton, Florida, p 133

Rogers D, Unal GG, Williams DJ et al 1979 Crystal structure of 3-epicaryoptin and the reversal of the currently accepted absolute configuration of clerodin. J Chem Soc Chem Commun 3:97–99

Sim GA, Hamor TA, Robertson JM 1962 The structure of clerodin: X-ray analysis of clerodin bromolactone. J Chem Soc 4133–4145

Simmonds MSJ, Blaney WM, Ley SV, Savona G, Bruno M, Rodriguez B 1989 The antifeedant activity of clerodane diterpenoids from Teucrium. Phytochemistry 28:1069–1071

Van Rie J, McGaughey WH, Johnson DE, Barnett BD, Van Mellaert H 1989 Mechanism of insect resistance to the microbial insecticide *Bacillus thuringiensis*. Science (Wash DC) 247:72–74

Wing KD, Slawecki RA, Carlson JR 1988 RH-5849, a nonsteroidal ecdysone agonist –effects on larval Lepidoptera. Science (Wash DC) 241:470–472

Chemical synthesis of bioactive polyamines from solanaceous plants

Adul Thiengchanya, Charoen Eung, Nongluk Tanikkul and Kan Chantrapromma

Department of Chemistry, Faculty of Science, Prince of Songkla University, Hat-Yai, 90112, Thailand

Abstract. Among the biologically active compounds isolated from medicinal plants are polyamine derivatives that show antihypertensive or antitumour activity. We have used previously available reagents to modify selectively spermidine, homospermidine and spermine backbones. Chemical synthesis of several naturally occurring derivatives has been achieved by simple, efficient procedures that can be used to produce high yields. This has facilitated evaluation of the structure–activity relationships of polyamines.

1990 Bioactive compounds from plants. Wiley, Chichester (Ciba Foundation Symposium 154) p 99–111

The polyamines putrescine, spermidine and spermine are widely distributed in biological systems, as free bases and also as alkyated or acylated conjugates with sugars, steroids, fatty acids and peptides (Ganem 1982). Many of these more elaborate structures exhibit notable biochemical and pharmacological properties (Ganem 1982). In recent years, there has been striking progress in the development of efficient methods for the formation of polyamine derivatives. There have been several reports (Bergeron et al 1982, Ganem 1982, Ganem & Tice 1983, Kunesch et al 1989, Sundaramoorthi et al 1984) of new procedures for the selective synthesis of derivatives of putrescine, spermidine and spermine.

Because of our interest in the synthesis of biologically active polyamine derivatives isolated from medicinal plants, we have tried to develop reagents for selective modification of specific nitrogens on the spermidine, homosperm-idines and spermine backbones. The idea was to build on what was already known. Reagents currently available for polyamine functionalization include a number of nitriles (Israel et al 1964), di-*tert*-butoxycarbonylated spermidines (Lurdes et al 1987), a cyclic urea (Chantrapromma et al 1980a) and hexahydropyrimidines (Chantrapromma et al 1980b). In the studies on hexahydropyrimidines **1**, some naturally occurring compounds, e.g. thermospermine **2** (Chantrapromma et al 1980b) and maytenine **3** (McManis & Ganem 1980), have been prepared.

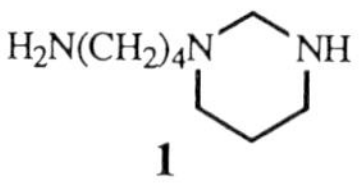

$$H_2N(CH_2)_4N\text{—}NH$$

1

$$H_2N(CH_2)_3NH(CH_2)_3NH(CH_2)_4NH_2$$

2

$$PhCH{=}CHCONH(CH_2)_3NH(CH_2)_4NHCOCH{=}CHPh$$

3

In this paper we describe the chemical modification of spermidine reagents and homologues and mono-BOC-hexahydropyrimidines **4** (Table 1), whereby selective functionalization of triamines with one, two or three different groups is achieved. This allows, for example, selective functionalization in any order with any three different acylating agents (Schemes 1, 2 and 3).

TABLE 1   Reaction of hexahydropyrimidine 1 with various protecting reagents

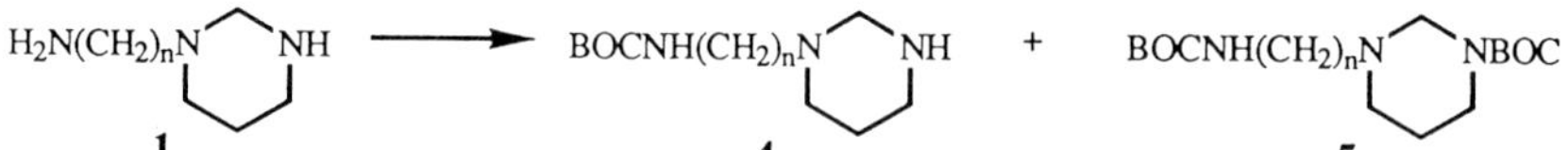

$$H_2N(CH_2)_nN\text{—}NH \longrightarrow BOCNH(CH_2)_nN\text{—}NH \quad + \quad BOCNH(CH_2)_nN\text{—}NBOC$$

1   **4**   **5**

n = 4,5 and 6

Reagents	Conditions	Ratios		% yield
		Mono	Di	
Di-*tert*-butyl dicarbonate, (BOC)₂O	RT, THF, 12 h	1	1	93
	0 °C, THF, 12 h	1	2	87
2-*tert*-butoxycarbonyloxymino)-2-phenylacetonitrile, BOC-ON	RT, THF, 12 h	1	1	85
	0 °C, THF, 12 h	5	1	81
9-Fluorenylmethyl chloroformate, FMOC-Cl	0 °C, THF, 12 h	1	2	75
2,2,2-Trichloroethyl chloroformate	0 °C, THF, 12 h	1	4	87
β,β,β-Trichloro-*tert*-butyl chloroformate	0 °C, THF, 12 h	1	2	85
S-BOC-2-mercapto-4,6-dimethylpyrimidine	0 °C, THF, 12 h	4	1	81
N-(*Tert*-butoxycarbonyloxy)-succinimide, BOC-OSU	0 °C, THF, 12 h	1	7	82

RT, room temperature; THF, tetrahydrofuran.

Scheme 1

$$BOCNH(CH_2)_nN \overset{\frown}{\underset{\smile}{}} NH \longrightarrow BOCNH(CH_2)_nN \overset{\frown}{\underset{\smile}{}} NCOR$$

4

$$\longrightarrow BOCNH(CH_2)_nNH(CH_2)_3NHCOR \longrightarrow NH_2(CH_2)_nNH(CH_2)_3NHCOR$$

Scheme 2

$$BOCNH(CH_2)_nN \overset{\frown}{\underset{\smile}{}} NH \longrightarrow BOCNH(CH_2)_nN \overset{\frown}{\underset{\smile}{}} NTCBOC$$

$$\longrightarrow H_2N(CH_2)_nN \overset{\frown}{\underset{\smile}{}} NTCBOC \longrightarrow RCOHN(CH_2)_nN \overset{\frown}{\underset{\smile}{}} NTCBOC$$

$$\longrightarrow TCBOCNH(CH_2)_3NH(CH_2)_nNHCOR \longrightarrow H_2N(CH_2)_3NH(CH_2)_nNHCOR$$

Scheme 3

$$\longrightarrow BOCNH(CH_2)_nN \overset{\frown}{\underset{\smile}{}} NBOC \longrightarrow BOCNH(CH_2)_3NH(CH_2)_nNHBOC$$

$$\overset{COR}{\longrightarrow BOCNH(CH_2)_3\overset{|}{N}(CH_2)_nNHBOC} \longrightarrow H_2N(CH_2)_3\overset{|}{N}(CH_2)_nNH_2$$

This methodology has been extended to spermine by simply coupling 2 moles of formalin with spermine to produce the corresponding *bis*-hexahydropyrimidine **6**. Using the protected spermine **6**, we were able to prepare a large number of kukoamine A derivatives in high yields (Schemes 4 and 5).

Scheme 4

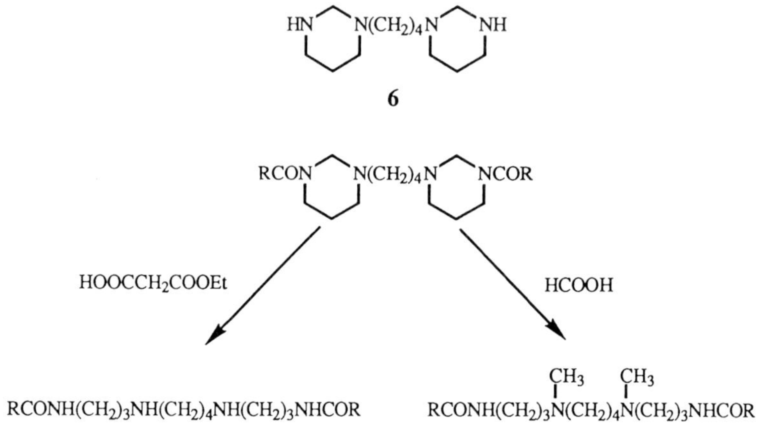

$$\text{HN}\underset{}{\overset{}{\diagdown}}\text{N(CH}_2)_4\text{N}\underset{}{\overset{}{\diagdown}}\text{NH}$$

6

$$\text{RCON}\underset{}{\overset{}{\diagdown}}\text{N(CH}_2)_4\text{N}\underset{}{\overset{}{\diagdown}}\text{NCOR}$$

HOOCCH$_2$COOEt   HCOOH

RCONH(CH$_2$)$_3$NH(CH$_2$)$_4$NH(CH$_2$)$_3$NHCOR

$$\text{RCONH(CH}_2)_3\overset{\overset{\text{CH}_3}{|}}{\text{N}}\text{(CH}_2)_4\overset{\overset{\text{CH}_3}{|}}{\text{N}}\text{(CH}_2)_3\text{NHCOR}$$

Scheme 5

$$\text{BOCN}\underset{}{\overset{}{\diagdown}}\text{N(CH}_2)_4\text{N}\underset{}{\overset{}{\diagdown}}\text{NBOC} \longrightarrow \text{BOCNH(CH}_2)_3\text{NH(CH}_2)_4\text{NH(CH}_2)_3\text{NHBOC}$$

$$\text{BOCNH(CH}_2)_3\overset{\overset{\text{COR}}{|}}{\text{N}}\text{(CH}_2)_4\overset{\overset{\text{COR}}{|}}{\text{N}}\text{(CH}_2)_3\text{NHBOC}$$

$$\text{H}_2\text{N(CH}_2)_3\overset{\overset{\text{COR}}{|}}{\text{N}}\text{(CH}_2)_4\overset{\overset{\text{COR}}{|}}{\text{N}}\text{(CH}_2)_3\text{NH}_2$$

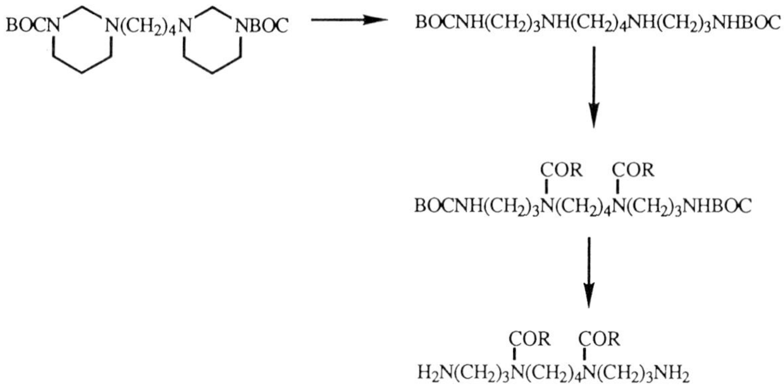

The above examples show that the hexahydropyrimidines **4** and **6** are latent sources of protected spermidines and protected spermine for use in chemical synthesis. Naturally occurring polyamine alkaloids include a large number of compounds in which the spermidine and spermine residues form part of the backbone. We therefore chose several representatives of this group of bioactive compounds as objectives for the synthetic studies based on hexahydropyrimidines.

Scheme 9

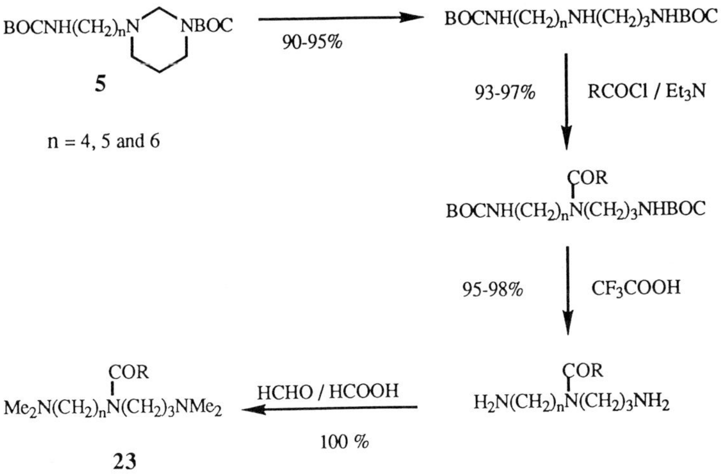

$$BOCNH(CH_2)_nN\text{<hexahydropyrimidine>}NBOC \xrightarrow[90\text{-}95\%]{} BOCNH(CH_2)_nNH(CH_2)_3NHBOC$$

5

n = 4, 5 and 6

93-97% | RCOCl / Et$_3$N

$$\underset{BOCNH(CH_2)_nN(CH_2)_3NHBOC}{\overset{\overset{\displaystyle COR}{|}}{}}$$

95-98% | CF$_3$COOH

$$\underset{Me_2N(CH_2)_nN(CH_2)_3NMe_2}{\overset{\overset{\displaystyle COR}{|}}{}} \xleftarrow[100\%]{HCHO \,/\, HCOOH} \underset{H_2N(CH_2)_nN(CH_2)_3NH_2}{\overset{\overset{\displaystyle COR}{|}}{}}$$

23

Kukoamine A derivatives 24 and 25

Our first demonstration of the use of bis-hexahydropyrimidine **6** in the formation of a naturally occurring spermine alkaloid was the selective synthesis of kukoamine A **24a** (Chantrapromma & Ganem 1981), an antihypertensive compound isolated from *Lycium chinense* (Solanaceae) (Funayama et al 1980). A number of kukoamine A derivatives **24b–g** and **25a–g** have been prepared in high yields. This work has made possible an extensive evaluation of structure–activity relationships in kukoamine A.

Conclusion

There are now many reagents available for the selective functionalization of polyamine nitrogens. Through the conversion of polyamines into their hexahydropyrimidine derivatives, we have been able to prepare a large number of acylated and alkylated polyamines in high yield. This work has made possible an extensive evaluation of structure–activity relationships of polyamine derivatives.

Acknowledgement

Generous financial support was provided by the Chulabhorn Research Institute.

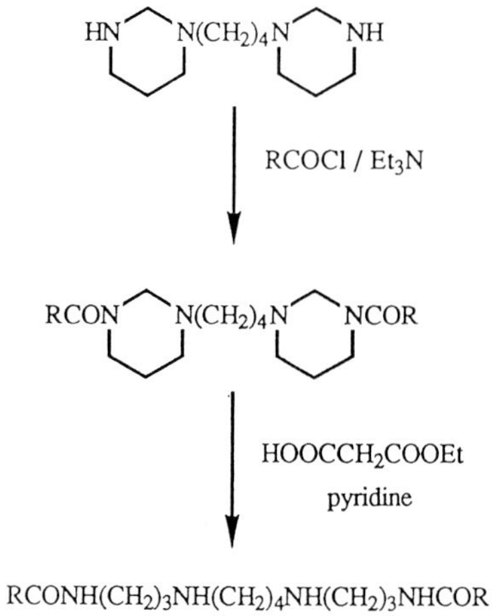

HN‿N(CH₂)₄N‿NH

$$\downarrow \text{RCOCl / Et}_3\text{N}$$

RCON‿N(CH₂)₄N‿NCOR

$$\downarrow \begin{array}{c}\text{HOOCCH}_2\text{COOEt}\\ \text{pyridine}\end{array}$$

RCONH(CH₂)₃NH(CH₂)₄NH(CH₂)₃NHCOR

24

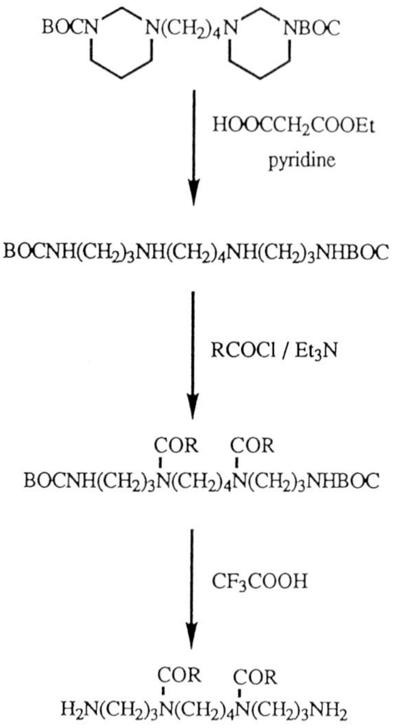

a R = CH₂CH₂(C₆H₃)(OH)₂ (Kukoamine A)

b R = CH₃

c R = Ph

d R = CH=CHPh

e R = CH₂CH₂Ph

f R = CH=CH(C₆H₄)(OH)

g R = CH=CH(C₆H₃)(OH)₂

BOCN‿N(CH₂)₄N‿NBOC

$$\downarrow \begin{array}{c}\text{HOOCCH}_2\text{COOEt}\\ \text{pyridine}\end{array}$$

BOCNH(CH₂)₃NH(CH₂)₄NH(CH₂)₃NHBOC

$$\downarrow \text{RCOCl / Et}_3\text{N}$$

$$\begin{array}{c}\quad\ \text{COR}\quad\ \text{COR}\\ \text{BOCNH(CH}_2)_3\text{N(CH}_2)_4\text{N(CH}_2)_3\text{NHBOC}\end{array}$$

$$\downarrow \text{CF}_3\text{COOH}$$

$$\begin{array}{c}\quad\ \text{COR}\quad\ \text{COR}\\ \text{H}_2\text{N(CH}_2)_3\text{N(CH}_2)_4\text{N(CH}_2)_3\text{NH}_2\end{array}$$

25

References

Bergeron RJ, Stolowich NJ, Porter CW 1982 Reagents for the selective secondary *N*-acylation of linear triamines. Synthesis 8:689–692

Chantrapromma K, Ganem B 1981 Total synthesis of kukoamine A, an hypotensive constituent of *Lycium chinense*. Tetrahedron Lett 22:23–24

Chantrapromma K, McManis JS, Ganem B 1980a Synthesis of cytotoxic spermidine metabolites from the solf coral *Sinularia brongersmai*. Tetrahedron Lett 21:2605–2608

Chantrapromma K, McManis JS, Ganem B 1980b The chemistry of naturally occurring polyamines. 2. A total synthesis of thermospermine. Tetrahedron Lett 21:2475–2476

Evans WC, Ghani A, Wooley VA 1972 Alkaloids of *Cyphomandra betacea* Sendtn. J. Chem Soc Perkin Trans I, p 2017–2019

Evans WC, Somanabandhu A 1977 Bases from roots of *Solanum carolinense*. Phytochemistry (Oxf) 16:1859–1860

Funayama S, Yoshida K, Konno C, Hikino H 1980 Structure of kukoamine A, a hypotensive principle of *Lycium chinense* root barks. Tetrahedron Lett 21:1355–1356

Ganem B 1982 New chemistry of naturally occurring polyamines. Acc Chem Res 15:290–298

Ganem B, Tice CM 1983 The chemistry of naturally occurring polyamines. 6. Efficient syntheses of N^1- and N^8-acetylspermidine. J Org Chem 48:2106–2108

Israel M, Rosenfield JS, Modest EJ 1964 Analogs of spermine and spermidine. I. Synthesis of polymethylenepolyamines by reduction of cyanoethylated, α,ω-alkylenediamines. J Med Chem 7:710–716

Kunesch G, Chuilon S, Ramiandrasoa F, Milat ML 1989 A new regioselective synthesis of N^1- and N^8-monoacylated spermidines. Tetrahedron Lett 30:1365–1368

Kupchan SM, Davies AP, Barboutis ST, Schnoes HK, Burlingame AL 1969 Tumor inhibitors. XLIII. Solapalmitine and solapalmitenine, two novel alkaloid tumor inhibitors from *Solanum tripartitum*. J Org Chem 34:3888–3893

Lurdes M, Almcida S, Grehn L, Ragnarsson U 1987 Selective protection of mixed primary-secondary amines. Simple preparation of N^1, N^8-bis (*t*-butoxycarbonyl) spermidine. J Chem Soc Chem Commun p 1250–1251

McManis JS, Ganem B 1980 Chemistry of naturally occurring polyamines. 1 Total synthesis of celacinnine, celabentine, and maytenine. J Org Chem 45:2041–2042

Sundaramoorthi R, Marazano C, Fourrey JL, Das BC 1984 Synthesis of N^4 acyl-spermidines. Tetrahedron Lett 25:3191–3194

DISCUSSION

Rickards: These compounds, particularly as their salts, would seem to be surface active agents and in that sense their biological activity might not be very specific. Is their activity specific in the tests you have done or do they have activity against a whole range of systems?

Chantrapromma: We have tested only kukoamine A for antihypertensive activity. The reason I made these analogues is to see whether these other derivatives of kukoamine A, spermine and spermidine, had greater anti-hypertensive activity.

Ley: The compounds certainly are surfactant-type materials but they are also potential metal-sequestering agents, particularly kukoamine, where there is the catechol unit which could bind metals very strongly. Do you know if kukoamine binds to metals? For example, can you use it as a crowning system? Is this what's happening in hypertension, where Ca^{2+} may be involved?

Chantrapromma: This is possible, but I have not looked at it.

Battersby: I was very interested by the ring opening of one of your reduced pyrimidine systems using formic acid, a sort of Eschweiler–Clarke reduction. Was that a specific reaction in one direction?

Chantrapromma: Yes, there is a 100% yield of the product of the *N*-methylation within the chain.

Fleet: I like the protecting group chemistry you did on the polyamines, but is it necessary? What happens if you just acylate the amines with no protection at all? Presumably, there is quite a lot of selectivity between the amines in the acylation of the unprotected amines?

Chantrapromma: It is necessary to use the protecting group to acylate the polyamines. Starting from spermidine or spermine you would get a number of compounds if you did not use a protecting group, and it would be very difficult to separate them.

Kuć: There is a lot of interest now in the role of polyamines in second messenger systems and signal transduction, in addition to the roles of G proteins and calmodulin in these systems. This might be a productive area for you to study.

Cragg: The N-alkylated derivatives have enhanced antitumour activity; have you tried preparing them?

Chantrapromma: Yes, we have tried to prepare a number of N-alkylated derivatives. We reduced the carbonyl groups to methylene groups and we are waiting for these compounds to be tested in bioassays. Unfortunately, we don't have any results on the relationship between activity and structure.

Liu: Have you tested the toxicity of your target compounds? Have you looked at bioactivity in relation to the structural modification?

Chantrapromma: I have not tested the cytotoxicity of these compounds because unfortunately there are not good facilities for this testing in our university. We hope to send our samples to someone who can do these tests. We have a very small lab in the pharmacology department working on the antihypertensive testing; we also have a lab for cytotoxic testing but we haven't done that yet.

Barz: Several cinnamoyl amides are involved in the synthesis of the plant cell wall, such as feruloyl tyramine or coumaroyl tyramine or spermidine derivatives. Do you have any evidence whether your compounds, such as the kukoamine, might also be involved in the synthesis of the plant cell wall?

Chantrapromma: No, I don't.

Fowler: I was very interested by your paper. Given the extensive literature on the Solanaceae in cell culture and tissue culture, have you tried to culture your plants?

Chantrapromma: We haven't tried this approach.

Fowler: It would be worth looking at this. The range of polyamine structures produced by cell cultures is quite wide and you may find some interesting precursors for modification from those systems. There may also be some key enzymes in there which may give you leads for alteration of active structures.

Yamada: In tissue culture, cells of higher plants produce polyamines, including spermidine, spermine and putrescine. These are rather toxic compounds and it is very difficult to accumulate this kind of polyamine compound in plant cells because of their toxic effects. The question is: how can we increase the amount of these compounds in cell cultures without damaging the cells?

Barz: Gonzales & Widholm (1985) treated tobacco cell suspension cultures with *p*-fluorphenylalanine and isolated a line that overproduces phenylalanine ammonia lyase. These cell cultures also overproduce the cinnamoyl amides in very large amounts, with the result that finally the cell cultures die—they over-produce themselves to death.

Fowler: Dr Kan, did you mention the total concentration of polyamines you found in your tissue? What sort of level of the dry weight of polyamines do you see in your plant extracts from the Solanaceae?

Chantrapromma: The total concentration of polyamines is very low. When the *Lycium chinense* plant (Kao Ki chai in the Thai language) was used as the source, it was difficult to detect kukoamine A by thin layer chromatography of the dried material; when we used fresh material we could see it but there was not much present. I want to synthesize the kukoamine A to get enough for testing, because the crude extract shows some antihypertensive activity.

Reference

Gonzales RA, Widholm JM 1985 Altered amino acid biosynthesis in amino acid analog and herbicide-resistant cells. In: Neumann KH et al (eds) Primary and secondary metabolism of plant cell cultures. Springer-Verlag, Berlin p337–343

Plagiarizing plants: amino sugars as a class of glycosidase inhibitors

George W. J. Fleet†, Linda E. Fellows° and Bryan Winchester*

†Dyson Perrins Laboratory, Oxford University, South Parks Road, Oxford OX1 3QY; °Jodrell Laboratory, Royal Botanic Gardens, Kew, Richmond, Surrey TW9 3DS, and *Department of Clinical Biochemistry, Institute of Child Health, University of London, 30, Guilford Street, London WC1N 1EH, UK

Abstract. Many polyhydroxylated alkaloids from plants are specific inhibitors of glycosidases. Information about these has led to the development of a wide range of both naturally occurring and synthetic inhibitors which may be used for mechanistic studies and for the purification of these enzymes. Sugar lactones are starting materials for highly efficient syntheses of deoxymannojirimycin and deoxy-fuconojirimycin and of a number of their iminoheptitol analogues. This has allowed an investigation of the relationship between mannosidase and fucosidase inhibition.

1990 Bioactive compounds from plants. Wiley, Chichester (Ciba Foundation Symposium 154) p 112–125

In the last ten years, a number of polyhydroxylated nitrogen heterocyclic compounds have been isolated from plants (Elbein 1987, Fellows & Fleet 1988), which are powerful and specific glycosidase inhibitors. In particular, a remarkable structural range of alkaloids has proved to inhibit exoglucosidases; these include pyrrolidines (such as DAB-1 [1,4-dideoxy-1,4-imino-D-arabinitol]), piperidines (such as deoxynojirimycin, homonojirimycin), indolizidines (such as castanospermine, 6-epicastanospermine) and pyrrolizidines (such as alexine, australine) (Fig. 1). Additionally, a number of natural products which are specific inhibitors of other glycosidases—particularly mannosidases—have been isolated, including swainsonine and deoxymannojirimycin. Analogous synthetic piperidines and pyrrolidines have been found to be powerful inhibitors of glycosidases, such as fucosidases and hexosaminidases (Fleet et al 1986). Such compounds have potential in the investigation and/or treatment of a number of different diseases, for example diabetes type 2, hereditary lysosomal storage diseases (Cenci di Bello et al 1983) including mannosidosis, fucosidosis, Tay Sachs disease, and certain cancers (Humphries et al 1986, Dennis 1986, Ostrander et al 1988) and various viral conditions.

Derivatives of compounds such as castanospermines, alexines and homo-nojirimycin are highly challenging synthetic targets, because they possess five adjacent chiral centres and seven adjacent functional groups.

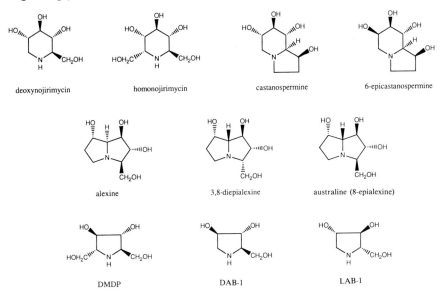

deoxynojirimycin   homonojirimycin   castanospermine   6-epicastanospermine

alexine   3,8-diepialexine   australine (8-epialexine)

DMDP   DAB-1   LAB-1

FIG. 1. Glucosidase inhibitors. LAB-1, 1,4-dideoxy-1,4-imino-L-arabinitol; DMDP, dihydroxymethyl-dihydroxypyrrolidine.

For most glycosidases there are many enzymes, even within a single organism, that hydrolyse the same glycosidic bond; thus, some inhibitors block the activity of one mannosidase efficiently but that of another only weakly. Substrate specificity is also important; inhibition of the hydrolysis of p-nitrophenyl mannopyranoside by Jack bean α-mannosidase may be a cheap and easy assay, but success or otherwise of such an inhibition is irrelevant if the object of the work is to inhibit a mannosidase of glycoprotein processing. For example, deoxymannojirimycin weakly inhibits the hydrolysis of the synthetic substrate, but powerfully inhibits that of the natural substrate by a mannosidase involved in processing glycoproteins. The ability to target a particular mannosidase is a desirable feature of a compound designed with a single biological objective (Cenci di Bello et al 1989). Swainsonine has potential for the treatment of cancer metastasis by inhibition of a mannosidase of glycoprotein processing, but it is also a very powerful inhibitor of lysosomal mannosidase. It is important to discover whether it is possible to design synthetic inhibitors for enzymes such as glycoprotein mannosidases even though the structures of the enzymes are unknown (Carpenter et al 1989).

Similarly, the antiviral properties of castanospermine and deoxynojirimycin derivatives are currently ascribed to the inhibition of glucosidase I (Walker et al 1987, Tyms et al 1987, Gruters et al 1987, Fleet et al 1988a). However, castanospermine is a potent inhibitor of gut disaccharidases, such as sucrase. Thus, specificity for inhibition of glucosidase I, without inhibition of sucrase, would be highly desirable (Sunkara et al 1989).

Synthetic strategy

Almost all the syntheses of the amino sugar glycosidase inhibitors start from homochiral materials, and the majority use sugars as starting materials. There are good reasons for this—the combination of adjacent functionality and chirality would require exceptional ingenuity for an asymmetric synthesis to compete with a starting material already possessing many of the features required in the target. Additionally, an important feature of any work on the relative ability of a set of stereoisomers to inhibit a given enzyme is confidence in the reliability of a result. If an inhibitor has a K_i of 10^{-9} M for the inhibition of a particular enzyme, then one part in ten thousand of such a compound may give a K_i of 10^{-5} M for an inert diastereomer.

There are also cases where even enantiomers are powerful inhibitors of related glycosidases; DAB-1 and the synthetic LAB-1 (1,4-dideoxy-1,4-imino-L-arabinitol) both inhibit glucosidases, although some α-glucosidases are much more susceptible to inhibition by one enantiomer (Scofield et al 1986). Thus in exploratory work, it is certainly safer, and probably more efficient, to consider the homochiral pool as the most desirable starting point. Also, if stereochemical ambiguities arise in a synthesis, it is much better for this to happen early, rather than late, so that there are more opportunities to separate the diastereomers.

Synthesis of deoxyfuconojirimycin and deoxymannojirimycin

Many of the problems in the synthesis of amino sugar glycosidase inhibitors can be illustrated by considering the synthesis of deoxyfuconojirimycin 1 and deoxymannojirimycin 2 and related compounds. All the targets have the same relative and absolute configurations at three of the adjacent carbons in a piperidine ring. Deoxyfuconojirimycin, a synthetic nitrogen analogue of L-fucose, is a very powerful inhibitor of several α-L-fucosidases and is the most potent and specific of all the amino sugar glycosidase inhibitors reported. The early syntheses of deoxyfuconojirimycin started from glucose, connecting C1 and C5 of the sugar by nitrogen (Fleet et al 1985). Although enough material could be prepared by this route to evaluate the compound, the number of steps makes this approach unattractive for the synthesis of large amounts. In contrast, a synthesis from D-lyxonolactone 3 is only six steps long and uses a single isopropylidene protecting group (Fleet et al 1989a). Treatment of lyxonolactone with acetone gives the 2,3-acetonide 4 in which only the C5 hydroxyl group remains free, allowing easy progress to the azidolactone 5. Addition of methyl lithium produces an azidolactol which on hydrogenation gives only the required epimer 6; deprotection of 6 by aqueous acid gives deoxyfuconojirimycin. This method can be used to synthesize large quantities of the inhibitor cheaply.

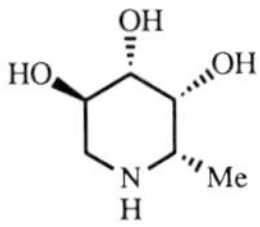

1

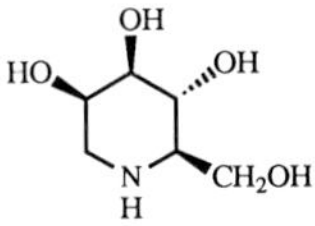

2

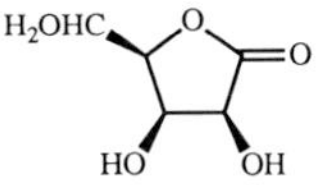

3

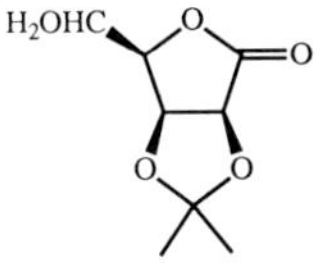

4

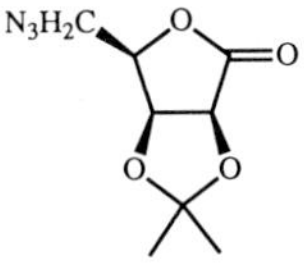

5

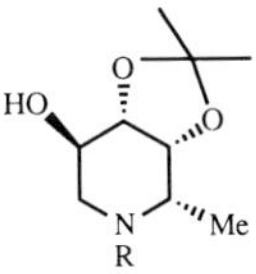

6

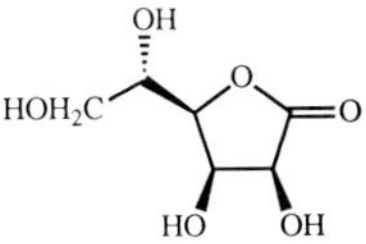

7

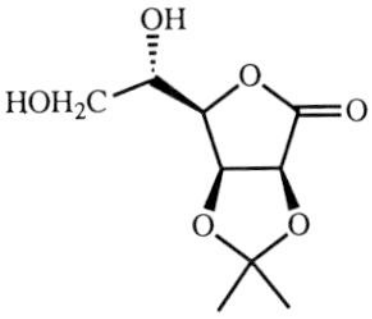

8

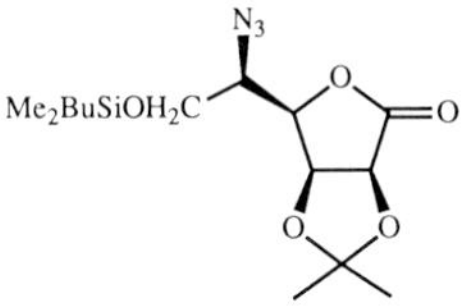

9

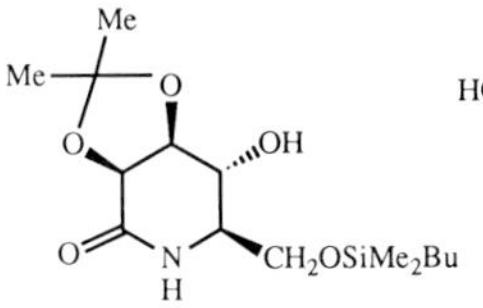

10

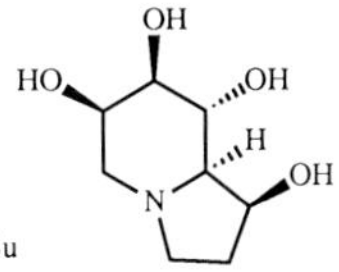

11

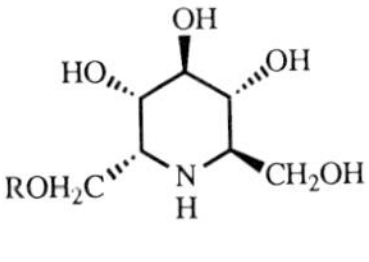

12 R = H
13 R = β-O-glucopyranosyl

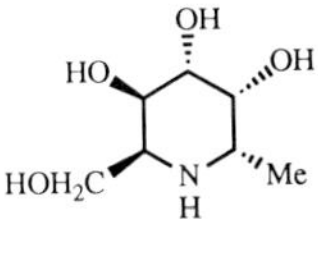

14

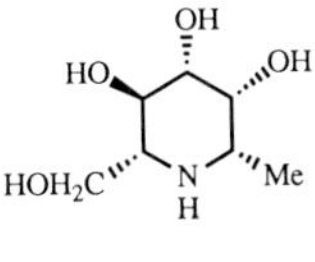

15

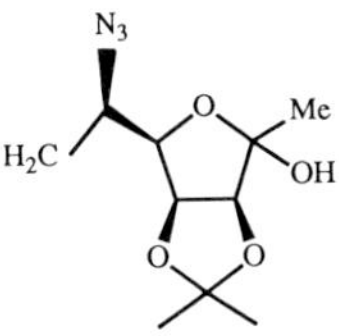

16

A similar strategy was used in the synthesis of deoxymannojirimycin starting from L-gulonolactone **7** (Fleet et al 1988b, 1989b). Formation of the 2,3-acetonide **8** and subsequent protection of the C6 hydroxyl function allowed introduction of an azide at C5 with inversion of configuration. Hydrogenation of the azidolactone **9** gave the δ-lactam **10**, which on reduction with borane and subsequent deprotection afforded deoxymannojirimycin in eight steps with an overall yield of 25%. The lactam **10** has also been used as an intermediate in the synthesis of 6-epicastanospermine **11**, first isolated as an oil from the mother liquors of the extraction of castanospermine (Molyneux et al 1986). This synthesis (Fleet et al 1988c) of 6-epicastanospermine is suitable for producing only small amounts of material; however, the route establishes the absolute stereochemistry while the relative stereochemistry has been determined by X-ray crystallographic analysis of the hydrochloride (Nash et al 1990). 6-Epicastanospermine is quite a potent inhibitor of a number of α-glucosidases, but has essentially no effect on lysosomal α-mannosidases, in spite of having a piperidine ring distinctly reminiscent of mannose.

Iminoheptitols

Azapyranose analogues of sugars lack an anomeric substituent, whereas the corresponding 2,6-dideoxy-2,6-iminoheptitols may allow syntheses of analogues of di- and oligosaccharides which take account of the geometry of the saccharide link. α-Homonojirimycin **12**, recently isolated from *Omphalea diandra* L.

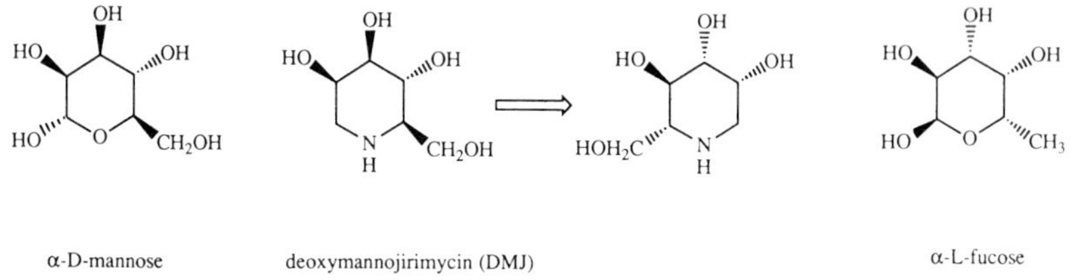

FIG. 2.   Structural relations between mannosidase and fucosidase inhibitors and the cognate sugars.

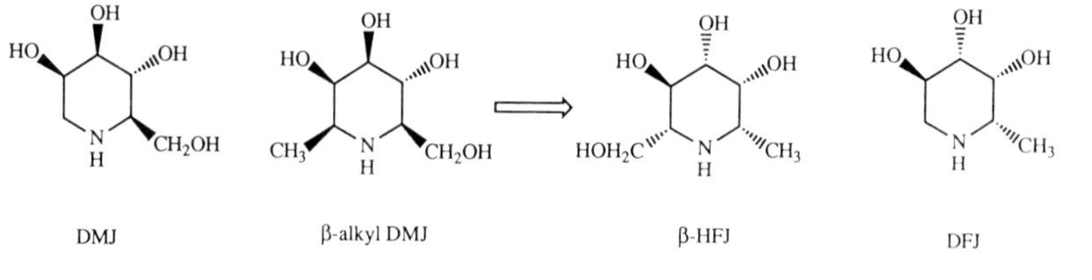

FIG. 3.   Structural relationship of deoxymannojirimycin (DMJ) to β-homofuconojirimycin (β-HFJ). β-HFJ is a strong, specific inhibitor of fucosidases; it does not inhibit mannosidases. DFJ, deoxyfuconojirimycin.

(Kite et al 1988), is the first example of a naturally occurring azapyranose analogue of such a heptitol. Iminoheptitols, such as α-homonojirimycin, may constitute a general class of glycosidase inhibitors in which the anomeric substituent confers additional potency and/or specificity in comparison to the corresponding azahexoses that lack such a substituent. For example, even before homonojirimycin had been isolated as a natural product, the β-glucopyranosyl derivative **13** (Anzeveno et al 1989) of α-homonojirimycin had been designed as a powerful α-glucosidase inhibitor. This compound is a drug candidate for antidiabetic therapy (Rhinehart et al 1987).

In an attempt to synthesize α-L-homofuconojirimycin (α-HFJ) **14**, the corresponding homologue of deoxyfuconojirimycin **1**, the epimeric β-homo-fuconojirimycin **15** was prepared. Treatment of the azido lactone **9** (a readily available intermediate in the synthesis of deoxymannojirimycin from L-gulonolactone) with methyl lithium produced the azidolactol **16**, which gave, after hydrogenation and deprotection, β-homofuconojirimycin **15**.

Mannosidase and fucosidase inhibition

Most of the polyhydroxylated piperidine compounds are highly specific inhibitors, but deoxymannojirimycin **2** is a stronger inhibitor of most fucosidases than it is of α-mannosidases (Evans et al 1985) (Fig. 2). This is not too surprising

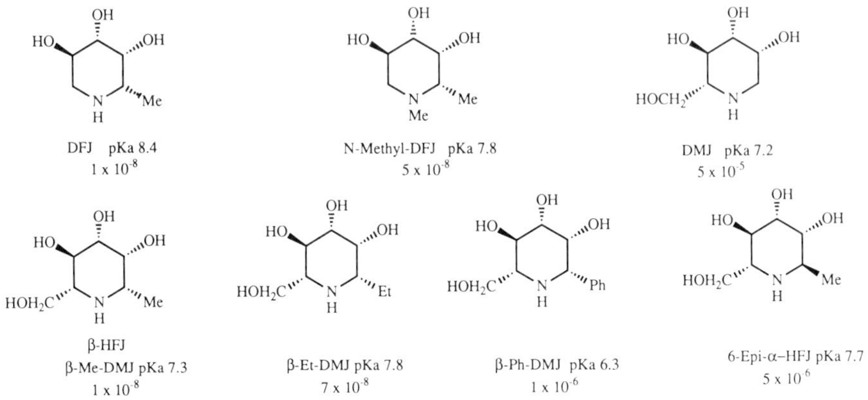

DFJ pKa 8.4
1 x 10^{-8}

N-Methyl-DFJ pKa 7.8
5 x 10^{-8}

DMJ pKa 7.2
5 x 10^{-5}

β-HFJ
β-Me-DMJ pKa 7.3
1 x 10^{-8}

β-Et-DMJ pKa 7.8
7 x 10^{-8}

β-Ph-DMJ pKa 6.3
1 x 10^{-6}

6-Epi-α–HFJ pKa 7.7
5 x 10^{-6}

FIG. 4. pK_a and K_i (M) for inhibition of hydrolysis of 4-umbelliferyl pyranoside catalysed by liver α-fucosidase. DFJ, deoxyfuconojirimycin; DMJ, deoxymannojirimycin; β-HFJ, β-homofuconojirimycin.

since it is essentially β-homofuconojirimycin lacking the 5-methyl group (Fig. 3). Several β-alkyl derivatives of deoxymannojirimycin were made from L-gulono-lactone (Fleet et al 1989c). All the C5 alkyl derivatives of deoxyfuconojirimycin were good competitive inhibitors of human liver and other α-fucosidases but only deoxymannojirimycin showed any appreciable inhibition of α-mannosidases. The common structural feature of all these compounds is the configuration of the secondary hydroxyl groups and it is tempting to speculate that this is the minimum structural feature necessary for inhibition of mammalian fucosidase (Winchester et al 1990). The pH dependency of fucosidase inhibition strongly supports the view that inhibition results from formation of an ion pair between the protonated inhibitor and a carboxylate group in the active site of the enzyme (Fig. 4).

α-Homomannojirimycin 17

Deoxymannojirimycin is a major biochemical tool for the investigation of mannosidases of glycoprotein processing but it was considered that α-homomannojirimycin 17 might be a more specific inhibitor of mannosidase because of the additional interaction of the anomeric substituent with the active site of mannosidases. The presence of an additional polar hydroxymethyl group in homomannojirimycin 17 relative to deoxymannojirimycin 2 implies that the former should be only a relatively weak fucosidase inhibitor (Table 1).

The synthesis of iminoheptitols with seven adjacent functional groups and five adjacent chiral centres represents a considerable challenge. All previous syntheses—and syntheses of alexines and castanospermines which also have five adjacent chiral centres—have started from a hexose and introduced the extra chiral centre late in the synthesis. An attractive alternative strategy is to start

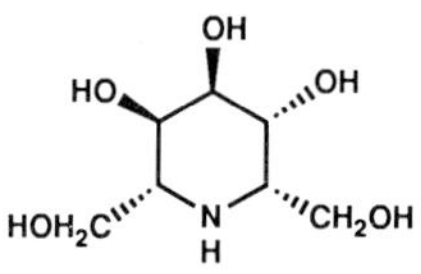

24

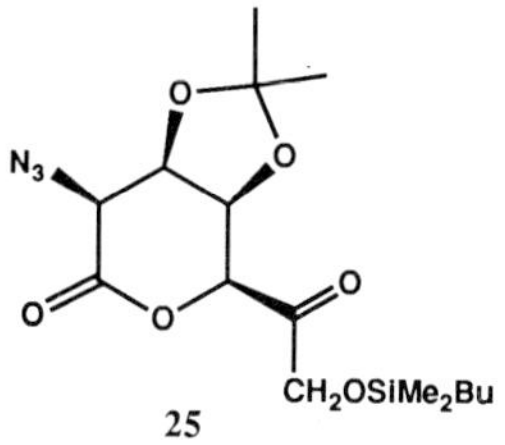

25

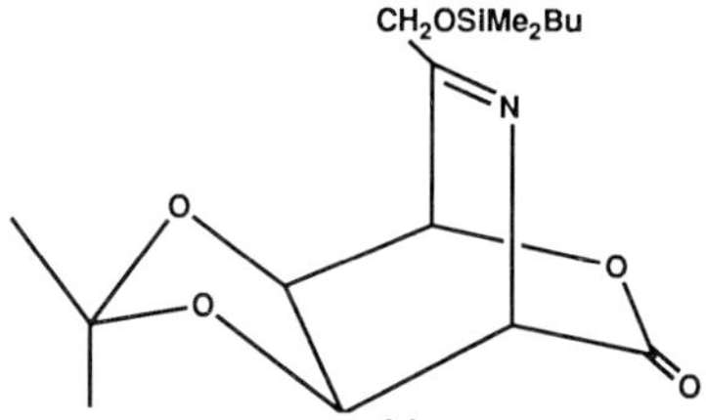

26

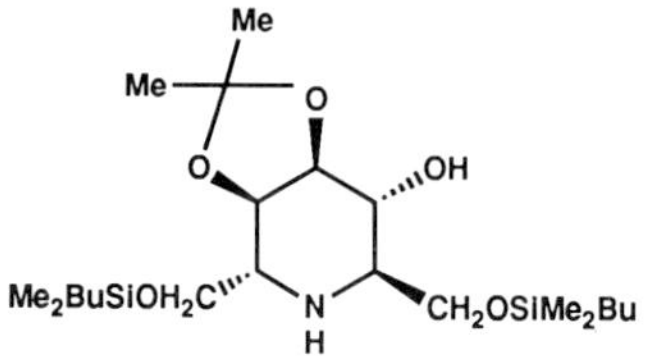

27

with a protected heptose derivative which already contains the required number of chiral centres. Diacetonides of heptonolactones, in which all the functionality other than a single hydroxyl group is protected by either acetone or the lactone function, are likely to be very useful intermediates for the synthesis of such complicated materials (Bruce et al 1990). For example, connection of C2 and C6 of a heptonolactone by nitrogen would lead to the synthesis of homonojirimycin derivatives.

In contrast to the behaviour of unprotected mannose, diacetone mannose gives the protected heptonolactones **18** and **19** in a ratio of 3:1. Although these isomers can be easily separated, *both* the triflates derived from **18** and **19** produce the same azide **20**, in which nitrogen has been introduced at C2, in excellent yield. Hydrolysis of the side chain acetonide in **20**, followed by protection of the primary group as a silyl ether, gives **21** in which only the secondary alcohol at C6 of the sugar is left unprotected. Hydrogenation of the azide allows the formation of the bicyclic amine **22** which, on reduction, gives the protected piperidine **23**; subsequent deprotection gives 6-epihomo-mannojirimycin **24**.

The ketone **25**, formed by oxidation of the secondary alcohol in **21**, undergoes an intramolecular aza-Wittig reaction on treatment with trimethyl phosphite to form **26**. Hydride reduction of **26** results in attack on the imine bond predominantly from the least hindered side to give **27**, with only traces of **23**. Deprotection of **27** by aqueous acid yielded α-homonojirimycin, which, in preliminary experiments, is about as good an inhibitor of mannosidases as is deoxymannojirimycin but shows no significant inhibition of fucosidase (Bruce et al 1989) (Fig. 5, Table 1). We are currently investigating the effect of α-homonojirimycin on processing mannosidases.

This is the first example of the use of such diacetonides of heptonolactones in the synthesis of such complex homochiral targets, but it is clear that easy access

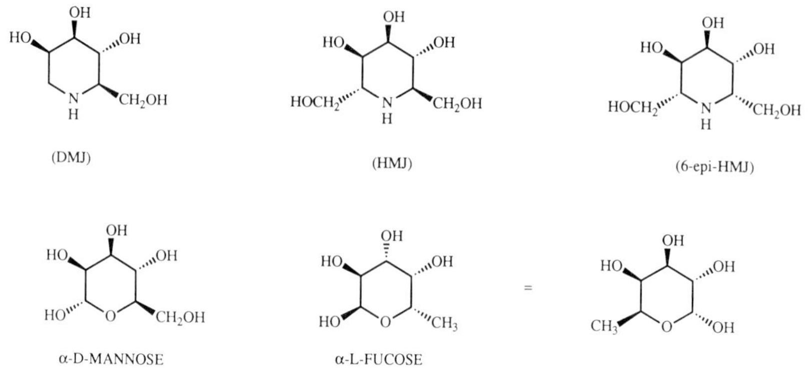

FIG. 5. Comparison of the structures of fucosidase and mannosidase inhibitors with those of the cognate sugars.

TABLE 1   Percentage inhibition of hydrolysis of 4-umbelliferyl pyranosides catalysed by liver α-fucosidase and α-mannosidase

| | α-Mannosidases | | | |
Inhibitor (1 mM)	Lysosomal	Golgi	Neutral	α-Fucosidase
Deoxymannojirimycin	58	45	21	91(K_i, 5 μM
Homomannojirimycin	49	56	30	29
6-Epihomomannojirimycin	0	0	0	96 (K_i, 4.5 μM

to a single hydroxyl group in such systems will allow the relatively easy synthesis of materials that would otherwise require extensive protection and deprotection.

The only published example of the inhibition of any enzymes other than hydrolases by this class of polyhydroxylated alkaloid is the inhibition of a glucosyl transferase by deoxynojirimycin (Newbrun et al 1983). It may well be that analogues of pyranosyl or furanosyl cations are inhibitors of other transformations involving sugars. Thus, only a small part of the range of biochemical transformations that may be affected by these natural and synthetic simple heterocyclic compounds may yet have been identified.

The ability of plant biochemists and synthetic organic chemists to work together has resulted in the development of a group of compounds which are powerful biochemical tools and which may have some chemotherapeutic potential.

Acknowledgements

The work (G. W. J. F. and B. W.) has been supported by SERC and by Monsanto G. D. Searle.

References

Anzeveno PB, Creemer LJ, Daniel JK, Kig C-HR, Liu PS 1989 2,6-Dideoxy-2,6-imino-7-O-β-D-glucopyranosyl-D-*glycero*-L-*gulo*-heptitol (MDL 25,637). J Org Chem 54:2539–2542

Bruce I, Fleet GWJ, Cenci di Bello I, Winchester B 1989 Iminoheptitols as glycosidase inhibitors: synthesis of, and mannosidase and fucosidase inhibition by, α-homomannojirimycin and 6-epi-homomannojirimycin. Tetrahedron Lett 30:7257–7261

Bruce I, Fleet GWJ, Girdhar A, Haraldsson M, Peach JM, Watkin DJ 1990 Retention and apparent inversion during azide displacement of α-triflates of 1,5-lactones. Tetrahedron 46:19–32

Carpenter NM, Fleet GWJ, Cenci di Bello I, Winchester B, Fellows LE, Nash RJ 1989 Synthesis of the mannosidase inhibitors, swainsonine and 1,4-dideoxy-1,4-imino-D-mannitol and of the ring contracted swainsonines, (1S, 2R, 7R, 7aR)-1,2,7-trihydroxypryrrolizidine and (1S, 2R, 7S, 7aR)-1,2,7-trihydroxypyrrolizidine. Tetrahedron Lett 30:7261–7264

Cenci di Bello I, Dorling P, Winchester B 1983 The storage products in genetic and swainsonine-induced human mannosidoses. Biochem J 215:693–696

Cenci di Bello I, Fleet G, Namgoong SK, Tadano K, Winchester B 1989 Inhibition of human α-mannosidases in vitro and in vivo by swainsonine analogues. Biochem J 259:855–861

Dennis JW 1986 Effects of swainsonine and polyinosinic:polycytidylic acid on murine tumour cell growth and metastasis. Cancer Res 46:5131–5136

Elbein AD 1987 Inhibitors of the biosynthesis and processing of N-linked oligosaccharide chains. Annu Rev Biochem 56:497–534

Evans SV, Fellows LE, Shing TKM, Fleet GWJ 1985 Glycosidase inhibition by plant alkaloids which are structural analogues of monosaccharides. Phytochemistry (Oxf) 24:1953–1956

Fellows LE, Fleet GWJ 1988 Alkaloidal glycosidase inhibitor from plants. In: Wagman GH, Cooper R (eds) Natural products isolation, chap 13. Elsevier Science Publishers, Amsterdam, p 539–559

Fleet GWJ, Shaw AN, Evans SV, Fellows LE 1985 Synthesis from D-glucose of 1,5-dideoxy-1,5-imino-L-fucitol, a potent α-L-fucosidase inhibitor. J Chem Soc Chem Commun p 841–842

Fleet GWJ, Smith PW, Nash RJ, Fellows LE, Parekh RB, Rademacher TW 1986 Synthesis of 2-acetamido-1,5-imino-1,2,5-trideoxy-D-mannitol and of 2-acetamido-1,5-imino-1,2,5-trideoxy-D-glucitol, a potent and specific inhibitor of a number of β-*N*-acetylglucosaminidases. Chem Lett p 1051–1054

Fleet GWJ, Karpas A, Dwek RA et al 1988a Inhibition of HIV replication by aminosugar derivatives. FEBS (Fed Eur Biochem Soc) Lett 237:128–132

Fleet GWJ, Ramsden NG, Witty DR 1988b Short syntheses of D-deoxymannojirimycin and D-mannonolactam from L-gulonolactone and of L-deoxymannojirimycin and L-mannonolactam from D-gulonolactone. Tetrahedron Lett 29:2871–2874

Fleet GWJ, Ramsden NG, Molyneux RJ, Jacob GS 1988c Synthesis of 6-epicastanospermine and 1,6-diepicastanospermine from L-gulonolactone and synthesis of L-6-epicastanospermine and L-1,6-diepicastanospermine from D-gulonolactone. Tetrahedron Lett 29:3603–3606

Fleet GWJ, Petursson S, Campbell A et al 1989a Short efficient synthesis of the α-L-fucosidase inhibitor, deoxyfuconojirimycin [1,5-dideoxy-1,5-imino-L-fucitol] from D-lyxonolactone. J Chem Soc Perkin Trans I p 665–666

Fleet GWJ, Ramsden NG, Witty DR 1989b Practical synthesis of deoxymannojirimycin and mannonolactam from L-gulonolactone. Synthesis of L-deoxymannojirimycin and L-mannonolactam from D-gulonolactone. Tetrahedron 45:319–326

Fleet GWJ, Namgoong SK, Barker C, Baines S, Jacob GS, Winchester B 1989c Iminoheptitols as glycosidase inhibitors: synthesis of and specific α-L-fucosidase inhibition by β-L-homofuconojirimycin and 1-β-C-substituted deoxymannojirimycins. Tetrahedron Lett 30:4439–4442

Gruters RA, Neefjes JJ, Tersmette M et al 1987 Interference with HIV-induced syncytium formation and viral infectivity by inhibitors of trimming glucosidase. Nature (Lond) 330:74–77

Humphries KJ, Matsumoto K, White S, Olden K 1986 Inhibition of experimental metastasis by castanospermine in mice: blockage of two distinct stages of tumour colonization by oligosaccharide processing inhibitors. Cancer Res 46:5215–5222

Kite GC, Fellows LE, Fleet GWJ, Liu PS, Scofield AS, Smith NG 1988 α-Homonojirimycin [2,6-dideoxy-2,6-imino-D-glycero-L-gulo-heptitol] from Omphalea diandra L.: isolation and glucosidase inhibition. Tetrahedron Lett 29:6483–6486

Molyneux RJ, Roitman JN, Dunnheim G, Szmilo T, Elbein AD 1986 6-Epicastanospermine, a novel indolizidine alkaloid that inhibits α-glucosidase. Arch Biochem Biophys 251:450–457

Nash RJ, Fellows LE, Girdhar A et al 1990 X-ray crystal structure of the hydrochloride of 6-epicastanospermine [(1S,6R,7R,8R,8aR)-1,6,7,8-tetrahydroxyoctahydroindolizine]. Phytochemistry 29:1356–1358

Newbrun E, Hoover CI, Walker GJ 1983 Inhibition by acarbose, nojirimycin and 1-deoxynojirimycin of glucosyl transferase produced by oral Streptococci. Arch Oral Biol 28:531–536

Ostrander GK, Scribner NK, Rohrschneider LR 1988 Inhibition of v-fms-induced tumor growth in nude mice by castanospermine. Cancer Res 48:1091–1094

Rhinehart BL, Robinson KM, Liu PS, Payne J, Wheatley ME, Wagner SR 1987 Castanospermine blocks the hyperglycaemic response to carbohydrate in vivo: a result of intestinal disaccharidase inhibition. J Pharmacol Exp Ther 241:915–920

Scofield AM, Fellows LE, Nash RJ, Fleet GWJ 1986 Inhibition of mammalian digestive disaccharidases by polyhydroxy alkaloids. Life Sci 39:645–650

Sunkara PS, Taylor DL, Kang MS et al 1989 Anti-HIV activity of castanospermine analogues. Lancet 1:1206–1207

Tyms AS, Berrie EM, Ryder TA et al 1987 Castanospermine and other plant alkaloid inhibitors of glucosidase activity block the growth of HIV. Lancet 2:1025–1026

Walker BD, Kowalski M, Goh WC et al 1987 Inhibition of human immunodeficiency virus syncytium formation and virus replication by castanospermine. Proc Natl Acad Sci USA 84:8120–8124

Winchester B, Barker C, Baines S, Jacob GS, Fleet G 1990 Inhibition of α-L-fucosidase by derivatives of deoxyfuconojirimycin and deoxymannojirimycin. Biochem J 265:277–282

DISCUSSION

Hall: In my own programme we are interested in the glycan residues in some of the seed proteins because they affect digestibility. There are many ways in which one can increase the usefulness of economic plants, one of which is to make them more digestible. In my own lecture I will talk about the removal of the asparagine to which the *N*-glycan is attached (Hall et al, this volume).

With respect to the targeting of materials to subcellular locations, since the trimming of sugar side chains presumably will only occur in specific places, for example once the protein has moved to the *cis*-Golgi, what is your feeling about the actual migration of molecules? Is it fortuitous where the processing occurs or is there some direction?

Fleet: I am not an expert in that. The enzymes responsible for the processing are probably membrane bound. Relatively high intracellular concentrations of the amino sugars arise, because the low pH inside the cells means that the amines are protonated.

Hall: Thank you, the importance of the glycans in functionality is clearly becoming much more recognized.

Battersby: George, you mentioned that you were getting beautifully crystalline hydrochlorides. I imagine that those are still highly soluble in water. What methods do you use to handle these substances?

Fleet: We got most of this expertise in the purification of the amino sugars from Linda Fellows and Robin Nash at Kew Gardens. If you dissolve these in water and add ethanol, these amines normally precipitate out. To make the hydrochlorides, we normally neutralize the amine, freeze dry it, put it back into water and precipitate the hydrochloride out with alcohol. In general these compounds are not soluble in anything that isn't mainly water.

Fowler: George, what is the specificity range of those glycosidases?

Fleet: Our collaborators have looked at a wide range of glycosidases. Our routine screens are against 14 human liver glycosidases, but a large number of these glycosidases have been compared with other mammalian glycosidases. Tony Scofield has studied mouse gut disaccharidases and Linda Fellows at Kew Gardens has looked at some plant and fungal enzymes; some of the compounds have also been studied by other investigators. In general, the enzymes are exoglycosidases and are relatively specific for the hydrolysis of one type of sugar link.

There are a large number of these compounds which inhibit α-glucosidase, but one inhibitor may inhibit one α-glucosidase better than another and *vice versa*. For example Tony Scofield showed that DAB-1 inhibits yeast α-glucosidase about 1000 times more effectively than does its mirror image LAB-1, but LAB-1 inhibits some mouse gut 1α-6-linked glucosidases, e.g. turanosidase, about 1000 times better than does DAB-1.

One of the problems with castanospermine as a potential inhibitor of glucosidase I of glycoprotein processing is its magnificent inhibition of sucrase (50% inhibition at about 10^{-9} M). There are real differences in the inhibition of different processing mannosidases by the different synthetic amino sugars. It is difficult to extrapolate from structural changes in a set of inhibitors of one glycosidase to the effects such changes may have on another enzyme. For example, the substitution of a methyl group in an inhibitor by an ethyl or even a phenyl group does not have much effect on human liver fucosidase, but it matters a lot to other fucosidases. These differences in enzyme specificities offer enormous potential for the synthesis of a compound that will selectively inhibit say glucosidase I without inhibiting other glucosidases.

Fowler: What intrigues me, George, is the use of those inhibitor systems to direct metabolic conversions. Could we use the relative specificities and inhibitory properties of these compounds to direct the flow of metabolite?

Fleet: There are considerable opportunities to affect mannosidase processing enzymes to different degrees with different amino sugar inhibitors.

Fowler: We have an interest in cloning human proteins into plant cell systems for production. One of the important problems there is glycosylation of the protein molecule for functionality. Do you see any use of your systems in selectively glycosylating or changing those proteins?

Fleet: Some of the inhibitors may allow selective changes in N-linked glycoproteins by virtue of their selective inhibition of individual processing enzymes, or by affecting the balance of general glycosidase versus transferase activity within the cell.

Ley: What is the role played in the biological mechanism by the NH group? Can it be replaced by, say, a CF_2 group, to make a cyclitol structure?

Fleet: Changing the ring nitrogen to a carbon atom creates other classes of glycosidase inhibitors which I cannot cover in this discussion. As far as the amino sugars are concerned, the protonated form of the amino sugars is probably the inhibitory form; the pKa of most of the amino sugars is around 7 or higher, so that under intracellular conditions they are almost completely protonated. It is difficult to predict the effect of nitrogen alkylation on the efficacy of inhibition. Dr Legler and Dr Bause in Köln looked at N-alkylation of deoxynojirimycin from methyl to about dodecyl and found that some of these alkylated materials are excellent glucosidase inhibitors. In contrast, N-alkylation of DAB-1 or LAB-1 eliminates almost all the glucosidase inhibition.

Ley: I can accept the polarity effect of a group or an NH being important, but other steric replacements and groupings that could do that are a CF or a CF_2 at that position.

Fleet: These types of modification would be interesting but the amount of chemistry involved increases dramatically. We have looked with Bryan Winchester at the effect on mannosidase inhibition by analogues of swainsonine. Changing the piperidine ring in swainsonine to a pyrrolidine ring effectively removes most of the ability to inhibit mannosidases. We have done extensive work on open chain analogues of swainsonine and it is not easy to explain the structure–activity relationships of the compounds.

Farnsworth: I assume all the biological activity you described is *in vitro*? Some of these compounds appear to be pretty toxic. Can you comment on this?

compounds and it is my contention that their biological activity, in particular their toxic effects on animals and microorganisms, is essential to the preservation of higher plants on this planet.

I take the view that these special metabolites have a broadly defensive role, that their synthesis and accumulation is beneficial and that their distribution is related to pressures from herbivores or the need to resist infection by microbes (Harborne 1988, 1989). Here, I shall focus on their role in chemical defence and show that many different strategies may be adopted within a plant to limit grazing. This explains why such apparently bewildering changes in secondary metabolism may occur in any given plant.

In this brief review, I shall concentrate on the plant and its vulnerability, rooted to the ground and unable to move, and suggest that the development of chemical barriers is one of the few avenues available for prevention of too much attention from herbivores. I shall say little about animal response, although clearly every chemical defence erected by a plant can eventually be overcome or circumvented by a hungry animal. There are at least six strategies by which the production of special metabolites is utilized for the overall benefit of plant life and these will now be considered in turn.

Accumulation of toxin

A simple but effective chemical protection may be afforded by the accumulation of a toxin, or a mixture of toxins, early in development and the maintenance of significant levels in most tissues throughout the life of the plant. This strategy is apparent in the chicory, *Cichorium intybus*, which is protected by three sesquiterpene lactones: 8-deoxylactucin, lactucin and lactupicrin. These lactones taste intensely bitter to mammals and deter insects from feeding. Their effectiveness is reinforced by their secretion in the latex, present throughout the plant, which readily accumulates around damaged tissue. Latex is a significant deterrent in its own right, preventing certain types of insect, such as leaf miners, from feeding on the plant.

Analyses of the concentrations of the lactones in chicory (Rees & Harborne 1985) show that the levels rarely drop below the amount (0.20% dry weight) required to deter feeding by the locust, *Schistocerca gregaria* (Table 1). Other antifeedants, including the phenolic coumarin cichoriin, are also present and probably synergize with the lactones in the protection against insects.

The bitter-tasting lactones, which protect chicory, also accumulate in closely related wild *Lactuca* species. By contrast, the cultivated *Lactuca sativa*, or lettuce, lacks them and it is hardly surprising that this salad plant is heavily preyed on by slugs and snails. The lactones were bred out of the plant to remove the bitter taste. Indeed, one reason why cultivated crops are so susceptible to herbivores is the fact that the naturally protective chemicals have usually been lost during domestication.

TABLE 1 Accumulation of three sesquiterpene lactones in chicory plants during the growing season

Month of analysis	% dry weight of total sesquiterpene lactones in:			
	Roots	Base leaves	Mid-stem leaves	Top-stem leaves
March	0.22[a]	0.06	—	—
April	0.38[a]	0.10	—	—
May	0.52[a]	0.21[a]	0.27[a]	—
June	0.81[a]	0.40[a]	0.32[a]	0.24[a]
July	0.11	0.17	0.22[a]	0.45[a]
August	0.25[a]	0.25[a]	0.15	0.38[a]
September	0.68[a]	0.22[a]	0.32[a]	0.28[a]

The lactones measured were 8-deoxylactucin, lactupicrin and lactucin.
[a]Concentrations which significantly deter feeding by locusts. Data from Rees & Harborne (1985).

The static type of defence in *Cichorium* and *Lactuca* may well occur in all sesquiterpene lactone-containing plants, although in some cases the lactones accumulate on the plant surface (in leaf wax or trichome) rather than within the tissue. Static defence may also operate generally in plants that accumulate terpenoids, although quantitative analyses have rarely been carried out to establish such an ecological role. Finally, even leaf-cutting ants may be deterred from utilizing a particular plant species, if terpenoid derivatives are present. For example, caryophyllene epoxide is the volatile repellent of *Melampodium divaricatum*. The ant *Atta cephalotes* avoids cutting the leaves of this plant, because the chemical damages the fungal symbiont present in the ant nest on which the survival of the ant depends (Hubbell et al 1983).

Induced accumulation of toxin

Maintenance of a relatively high level of secondary metabolite throughout the life cycle is arguably 'costly' to the plant. It requires synthesis from simple precursors, accumulation and transport to a site of storage or action. A better strategy would be to operate a secondary pathway at a relatively low rate and then 'turn on' high levels of the product in response to feeding by insects. This type of induced defence was first recognized about 1980 and has now been recorded in a variety of plants, although it is not a universal response. Plants which respond in this way to the trigger of insect feeding fall into two broad categories: trees, where the toxin which increases in amount is usually condensed and/or hydrolysable tannin (e.g. *Quercus rubra*); and herbs, where the toxin may be an alkaloid or a terpenoid. The response is rapid, comparable with the phytoalexin response, and may involve an increase in amount of 50% or more over the control.

The increase in alkaloid production in the wild tobacco plant, *Nicotiana sylvestris*, reaches a maximum 10 days after stimulation from feeding by larvae of the tobacco hornworm, *Manduca sexta*. Nicotine and nornicotine increase in amount to 220% of the control. Mechanical damage of the leaf can trigger a greater increase (up to 550% of the control) so that the magnitude of the response is theoretically greater than that measured in the field. The fact that attack by herbivores triggers only a fraction of the response to mechanical damage may reflect an adaptation of this particular insect to the plant (e.g. it may avoid cutting the veins of the leaf). Alkaloid is synthesized in the roots and is then transported to the upper part of the plant; any interference with root growth may reduce the magnitude of the response (Baldwin 1988).

This induced response in alkaloid-producing plants is likely to vary in degree in different species. We have evidence (J. B. Harborne & M. B. Khan, unpublished results) that in *Atropa acuminata* tropane alkaloid synthesis after mechanical damage reaches a maximum in a shorter time (eight days) than does

TABLE 2   Increases in alkaloid in *Atropa acuminata* leaves after mechanical damage

Day	% dry weight of total alkaloid	
	Control leaves	Damaged leaves[a]
0	0.157	0.155
1	0.150	0.182
2	0.160	0.195
4	0.163	0.216
6	0.162	0.217
8	0.157	0.234
10	0.174	0.194
12	0.167	0.188
14	0.164	0.178
16	0.158	0.169
18	0.162	0.175
20	0.166	0.172
22	0.170	0.185
24	0.174	0.185

The alkaloid comprised principally atropine, scopolamine and apoatropine. Values are average of three determinations; variations are within $\pm 0.01\%$.
[a]Leaves damaged on Day 0 by removal of approximately 50% of tissue from all leaves on the plant.

synthesis of nicotine in *Nicotiana* (Table 2). The maximum amount of total alkaloid in damaged tissue is 150% of the control. However, greater increases in tropane alkaloid levels in the leaf of *A. acuminata* can occur by limiting the nutrient status (e.g. lowering available potassium) of the plants. Plants may become more poisonous and unpalatable to herbivores simply as a result of their growing on poor soils.

Chemical protection of vulnerable tissues

Another defence strategy, present in the coffee plant, *Coffea arabica*, is the restriction of toxin synthesis to those tissues which are most vulnerable to attack and then only when they are susceptible to herbivores. Thus the purine alkaloid caffeine is produced in young soft coffee leaves in amounts equivalent to 4% of the dry weight. As the leaf thickens, toughens and matures, the rate of caffeine biosynthesis decreases exponentially from 17 to 0.016 mg/day per gram of leaf. Later, when the coffee bean is newly formed, the concentration of caffeine stands at about 2% of the dry weight, whereas the mature bean, where the seeds are surrounded by a solid protective endocarp, contains much lower amounts (0.24% dry weight). An additional advantage in this system is that the nitrogen-rich purine alkaloid is recycled for use in primary metabolism whenever it is not required for defence (Frischknecht et al 1986).

An even more dramatic example of chemical protection being concentrated in vulnerable plant organs is the paper birch, *Betula resinifera*. In this tree, juvenile growth-phase internodes are rendered unpalatable to the snowshoe hare, a major herbivore where the tree grows in Alaska, by an enormous concentration (up to 30% dry weight) of the triterpenoid, papyriferic acid. This is 25 times the level found in mature internodes, when such protection is less important. The triterpenoid is deposited as a resinous solid on the surface of the juvenile twigs and, combined with volatile oils in the resin, makes the juvenile tree highly deterrent to grazing by hares (Reichardt et al 1984). Feeding experiments with oatmeal containing 2% papyriferic acid showed that it is highly distasteful to the hare and rejection is based on the potential toxicity of this defence compound. Experiments with other food plants of the snowshoe hare indicate that vulnerable juvenile tissues are similarly protected by unique chemical deterrents: the stilbene pinosylvin in *Alnus crispa*, the monoterpene camphor in *Picea glauca* and 2,4,6-trihydroxydihydrochalcone in *Populus balsamifera* (Jogia et al 1989).

Yet another example of chemical protection of juvenile growth is in the holly plant, *Ilex opaca*, where the young leaves are defended from insect feeding by high levels of saponins. As the leaves mature and their structural defences are developed, there is a striking drop in saponin levels. In one example, on 30 April young leaves had 135 mg/g dry weight whereas on 11 June in the same year, this had fallen to 35 mg, a value maintained in older leaves (Potter & Kimmerer 1989).

Switch in chemistry during ontogeny

Another device by which insects and other herbivores may be successfully deterred from grazing is a switch in chemical defence within the plant so that later leaves have a different chemistry from the first leaves. Adaptation to such a complex defence system could pose a problem to many potential herbivores. Proksch et al (1990) have found that such a switch takes place in *Ageratina adenophora*, a member of the Compositae family.

Three insecticidal chromenes accumulate in leaves at nodes 1 to 8 in *Ageratina* but they are absent from leaves above node 8. Their place is taken by chlorogenic acid and a sesquiterpene diketone in all the upper leaves. Furthermore, these two new substances show either contact toxicity or growth-retarding effects on the larvae of an insect that feeds on a wide range of plants, the variegated cutworm *Peridroma saucia*. How widespread such changes in chemistry during ontogeny are is not yet known, but it is a fascinating aspect of secondary metabolism that deserves further exploration.

Hormonal protection

The subtle protection of plant tissues through the synthesis and accumulation of relatively massive amounts of insect hormones or hormone-mimics is now well established (cf. Harborne 1988). Numerous phytoecdysones, plant analogues of insect moulting hormones, have been recorded in plants and have a wide distribution, especially in gymnosperms and in ferns. These substances, when taken orally, have the ability to upset normal metamorphosis in test insects, causing sterility and death.

Likewise, juvenile hormone analogues and anti-juvenile hormones (e.g. the precocenes) have been characterized from a variety of plant species. The protective value of insect hormonal synthesis by plants has been underlined recently by the discovery of a true insect juvenile hormone, JHIII, in large amounts in leaves of the sedge, *Cyperus iria* (Toong et al 1988). The concentration in the plant (151 µg/g wet weight) is at least 150 times the largest amount that ever accumulates in insect tissue and is thus potentially able to disrupt normal development in any insect feeding on the plant. Indeed, experiments with the grasshopper *Melanoplus sanguinipes* showed that the JHIII in this sedge caused sterility in the female, so this represents a novel defensive mechanism which is probably effective against a wide range of insect species.

Variability in palatability within the plant

Leaves of long-lived plants, i.e. trees, are particularly vulnerable to herbivorous attack, since they are produced year after year, sometimes for centuries, and insect populations can 'home in' on a readily available and predictable food resource. Because insect predation has the potential to remove such a tree, there must be some innate resistance present and recent experiments indicate that this may be represented by variable defence. Thus, the patterns of insect grazing on trees indicate that a proportion of the leaves receive a low level of grazing, while chemical analyses show that the concentrations of defensive toxin (usually tannin) vary significantly from leaf to leaf, even on the same branch. Within-leaf variation in defensive chemistry places a significant constraint on the insect grazer and makes it more vulnerable to its predators, since it takes more time to find a feeding site. As Schultz (1983) puts it: 'the situation can be said to resemble a shell game in which a valuable resource (suitable leaves) is "hidden" among many other similar-appearing but unsuitable resources. The insect must sample many tissues to identify a good one'.

Such variable defence may apply to leaves on the same tree but it could also apply to leaves of different individuals within a population of trees. In the neo-tropical tree *Cecropia peltata*, leaves of some individuals are rich in tannin and low in herbivore damage, while leaves of other individuals are low in tannin and high in herbivore damage. The tannin levels in the leaves can vary from 13 to 58 mg/g dry weight (Coley 1986). There appears to be a significant cost to the tree in terms of tannin production so that high tannin trees produce fewer leaves than the low tannin trees, which hence can 'afford' to lose more leaves to herbivores. The presence of high tannin individuals in the population will presumably confer benefit on the low tannin members, since insects still have an unpredictable food resource (see above) and will have to spend time selecting low tannin trees for feeding.

In herbs, variations in the palatability of leaves are also apparent, notably in the case of cyanogenesis. Cyanogenic glycosides are bitter tasting and such plants are normally rejected, quite apart from their poisonous nature. The expression of cyanogenesis is under strict genetic control and there are significant variations in the frequency of cyanogenesis from population to population in many cyanogenic species. The protective value of cyanogenesis has been much debated, simply because it is so variable in nature. It is also costly to the plant in that cyanogenic glycosides are directly formed from amino acids used for protein synthesis. The sheer cost of production may explain why cyanogenesis became established as a variable defence in the first instance.

Recent experiments by Burgess & Ennos (1987) with slugs feeding on clover show that there are real advantages to the plant in maintaining cyanogenesis as a variable defence. Slug populations vary in their response to cyanogenesis according to the frequency of cyanogenic plants in a given population. Slugs from sites with a low frequency of cyanogenic clover avoid eating cyanogenic forms, whereas slugs from sites where cyanogenesis is frequent become adapted (via detoxification) and show little discrimination between the two forms.

In clover populations where the frequency of cyanogenesis is low (i.e. between 11 and 24%), cyanogenic forms will survive, since the slugs will graze exclusively on acyanogenic forms. By contrast, in clover populations where the frequency of cyanogenesis is high (65% and above) slugs will feed indiscriminately on both forms. In these plants, however, another factor comes into play: the cyanogenic glycosides are concentrated in the stem and cotyledon rather than in the leaf (Horrill & Richards 1986). So, although the cyanogenic forms are grazed, some will survive simply because the organs most vulnerable to predation are not eaten.

Selective feeding by slugs and snails (cf. Kakes 1989) will thus ensure that the frequency of cyanogenesis in any given population is liable to fluctuate from year to year and other environmental factors may also affect this frequency. As a food for molluscs, clover will continue to be an unpredictable resource and in the long term clover populations will benefit from the polymorphism of the cyanogenic character.

Conclusion

Plants have evolved a variety of defence strategies based mainly on secondary metabolism that allow them to avoid overgrazing by herbivores. No chemical defence is absolute and all barriers are capable of being breached by phytophagous insects. However, if several defences are present together in the same plant, few animals will be able to adapt to feeding on it. Bracken, *Pteridium aquilinum*, exemplifies a plant which is well protected from grazing (Table 3). The well-developed secondary chemistry of bracken must contribute to its biological success. In the United Kingdom, it is a weed that is growing out of control, because of its lack of predators (Lawton 1988).

Research on the role of secondary metabolites in plants is highly relevant to the utilization of some of these substances in medicine. Such research reveals likely biological activities and provides valuable information on production and turnover in the plant. Our own studies on tropane alkaloid synthesis in *A. acuminata* (Table 2) were initiated to determine the optimal environment for maximum alkaloid production in this plant. Information about the protective

TABLE 3   Bracken plant defences against being eaten

1. Five insect moulting hormones[a]
2. Thiaminase (vitamin destroyer)
3. Tannin[a] (anti-nutritional agent)
4. Cyanogens[a] (poisonous)
5. Silica and lignin (structural protection)
6. Ant nectaries (protection by ants)
7. Pterosins[a] (carcinogens)

[a] Products of secondary metabolism; for further information, see Jones (1983).

role of atropine against herbivores was incidental to that programme. Accurate quantitative data on secondary product synthesis are still scarce and ecological studies of the type described in this review are bound to add to our present imperfect knowledge of the bioactive compounds from plants.

References

Baldwin IT 1988 The alkaloidal responses of wild tobacco to real and simulated herbivory. Oecologia (Berl) 77:378–381

Burgess RSL, Ennos RA 1987 Selective grazing of acyanogenic white clover: variation in behaviour among populations of the slug *Deroceras reticulatum*. Oecologia (Berl) 73:432–435

Coley PD 1986 Costs and benefits of defence by tannins in a neotropical tree. Oecologia (Berl) 70:238–241

Frischknecht PM, Ulmer-Dufek J, Baumann TW 1986 Purine alkaloid formation in buds and developing leaflets of *Coffea arabica* : expression of an optimal defence strategy? Phytochemistry (Oxf) 25:613–616

Harborne JB 1988 Introduction to ecological biochemistry, 3rd edn. Academic Press, London

Harborne JB 1989 Recent advances in chemical ecology. Nat Prod Rep 6:85–109

Haslam E 1986 Secondary metabolism—fact and fiction. Nat Prod Rep 3:217–249

Horrill JC, Richards AJ 1986 Differential grazing by the mollusc *Arion hortensis* on cyanogenic and acyanogenic seedlings of the white clover *Trifolium repens*. Heredity 36:277–281

Hubbell SP, Wiemer DF, Adejore A 1983 Antifungal terpenoid defends a neotropical tree (*Hymenaea*) against attack by fungus-growing ants. Oecologia (Berl) 60:321–327

Jogia MK, Sinclair ARE, Andersen RJ 1989 An antifeedant in balsam poplar inhibits browsing by snowshoe hares. Oecologia (Berl) 79:189–192

Jones CG 1983 Phytochemical variation, colonisation and insect communities: the case of bracken fern. In: Denno RF, McClure MS (eds) Variable plants and herbivores in natural and managed systems. Academic Press, New York, p 513–558

Kakes P 1989 An analysis of the costs and benefits of the cyanogenic system in *Trifolium repens*. Theor Appl Genet 77:111–118

Lawton JH 1988 Biological control of bracken in Britain: constraints and opportunities. In: Wood RKS, May MJ (eds) Biological control of pests, pathogens and weeds. The Royal Society, London, p 225–244

Potter DA, Kimmerer TW 1989 Inhibition of herbivory on young holly leaves: evidence for the defensive role of saponins. Oecologia (Berl) 78:322–329

Proksch P, Wray V, Isman MB, Rahaus I 1990 Ontogenetic variation of biologically active natural products in *Ageratina adenophora*. Phytochemistry (Oxf) 29:453–458

Rees SB, Harborne JB 1985 The role of sesquiterpene lactones and phenolics in the chemical defence of the chicory plant. Phytochemistry (Oxf) 24:2225–2231

Reichardt PB, Bryant JP, Clausen TP, Wieland GD 1984 Defense of winter-dormant Alaska paper birch against snowshoe hares. Oecologia (Berl) 65:58–69

Schultz JC 1983 Impact of variable plant defensive chemistry on susceptibility of insects to natural enemies. In: Hedin PA (ed) Plant resistance to insects. American Chemical Society, Washington DC, p 37–54

Toong YC, Schooley DA, Baker FC 1988 Isolation of insect juvenile hormone III from a plant. Nature (Lond) 333:170–171

DISCUSSION

Hall: You promised to tell us a little about the volatile signal, which reputedly passes from one red oak tree to another, warning of herbivory. I would like to hear what hard evidence there is for the 'talking tree' hypothesis.

Harborne: Zeringue (1987) showed that in cotton, myrcene is released from the leaf of one plant and apparently triggers an increase in the synthesis of the gossypol-like compounds in the leaf of a second plant: these compounds are insect poisons. So in this type of lab experiment it is possible to show that a volatile signal passes between plants. Cotton is a herb and it's much easier to do such experiments with herbs.

The main problem with working with trees is the enormous difficulty of separating one tree from another. A lot of the criticism of the early work of Schultz & Baldwin (1982) was that the roots of the trees were intertwined, so the chemical message could have passed through the root rather than have been released as a volatile signal.

There are a number of volatile compounds given off by leaves, such as leaf alcohol (*E*-hex-2-en-1-ol), and ethylene, which are possible messengers. I agree that we still need definitive evidence that this type of signalling occurs in nature.

Battersby: You showed the marked decrease in caffeine content in both leaves and fruits of the coffee plant as they mature. Is it known what happens to the caffeine? Is it translocated or degraded, or is it just the increase in size of the organ that is responsible for the lower concentration?

Harborne: Caffeine is continually being produced and turned over. All you need to do is reduce the synthesis to cause the caffeine content to drop like that. It is really just turning off the synthesis. In the coffee plant the rate of caffeine synthesis decreases exponentially from 17 to 0.016 mg/day per g leaf tissue during leaf maturation. I am sure that caffeine is recycled, being so rich in nitrogen. I don't think anybody has identified the enzymes of turnover but I would be very surprised if that wasn't what was happening.

Yamada: Professor Harborne, you mentioned experiments with *Atropa acuminata* where damage causes an increase in the hyoscyamine content. Is this physical damage or another kind of damage?

Harborne: This is physical damage. You can either cut out a hole in the leaf with a punch or you can cut off half the leaf. Whether you cut the veins may also vary the effect on the alkaloid production.

One thing which spoils my story to some extent is that if you stress the plant by reducing the amount of K^+ available, you can get a higher concentration of alkaloid synthesis than after mechanical damage. It doesn't spoil my story from the point of view of protection from herbivores, but it does indicate that all sorts of environmental factors can modify secondary metabolism. In potassium deficiency, there tends to be a build-up of putrescine, which is a precursor for the tropane alkaloids. So you may just be providing a lot more precursor to alkaloid production.

Concerning the relationship between nitrogen content and alkaloid synthesis, you normally assume that an increase in nitrogen will lead to an increase in alkaloid. That is not necessarily so, because if you increase the amount of nitrogen available, you increase the productivity of the plant. The actual alkaloid production may be more but there is a lot more plant, so in relation to the biomass it may not increase. In fact if you reduce the amount of nitrogen, you can show a net increase in the amount of alkaloid dry weight under conditions where you would have thought that alkaloid synthesis would be repressed.

McDonald: I presume that defence mechanisms as you describe them have only been identified for a minority of plants. Do you believe that they are widespread throughout the plant kingdom?

Harborne: Yes, you are right to point out that this is based on a few experiments on a very limited number of plants. We know that secondary compounds are extremely widely distributed and almost every type of plant has some type of secondary compound accumulating in its tissues. If you take sesquiterpene lactones, there have been experiments on only three or four species in the Compositae that showed a protective effect. Some work has been done on *Vernonia* by Burnett et al (1987) but as far as I know our study on chicory has been the only deliberate attempt to establish a defensive role for these lactones.

MacMillan: It has been estimated that there are over 80 000 secondary plant metabolites, so there is a great deal of scope for the idea that they have a function in the plant kingdom.

Harborne: There are thousands of species of Compositae to examine and the range of known sesquiterpene lactones increases exponentially with time. There are at least 5000 known structures, largely from Professor Bohlmann's work, but there are many more still to be examined. Any particular plant may have 15 or 20 of these compounds and it's difficult not to think that they serve more than one function. Compounds certainly can be multifunctional and one of the problems in establishing a role is that you might be going up the wrong street— you might think it's protecting the plant from insects, but it's actually protecting it from something quite different.

Barz: There are a lot of confusing statements in the literature as to the exact role of secondary metabolites. If we really want to see the function of these compounds, we have to study the plant in its natural environment with the natural pathogens, which means doing field studies, not just laboratory studies.

Kuć: I would like to tell you of an adventure that we had with tobacco and the duvatrienediols (DVT). These are produced in the trichomes that cover the surface of tobacco leaves. There is a very serious disease of tobacco (blue mould) caused by a fungus, *Peronospora tabacina*.

We worked with burley tobacco which was available in Kentucky and was very susceptible to blue mould. As the plants aged in the greenhouse, they became resistant and often immune. We carried out experiments in the greenhouse and found that as the plants aged there was a remarkable increase in the amounts of the DVT diols which accumulated on the leaves. We could account for over 95% of the antifungal activity on the basis of these compounds. We could remove the diols without damaging the leaf, put them back on and it was a very clean-cut case. We did the experiments three times in the greenhouse and occasionally we would go into the field to collect tissue. Yet we knew that there were occasions when the disease was very serious in the fields but DVT levels were high.

We decided to monitor the levels of DVT and see whether we could predict when it would be advisable to spray plants with fungicides to avoid spraying unnecessarily. We did extensive field tests in eleven counties in Kentucky. We found all sorts of complications—disease would break out even when our analysis showed that the DVT were present in high concentrations. We finally figured this all out, and what it came down to was rain! The DVT have an appreciable water solubility. In the greenhouse we water our pots by watering the soil surface. The fungus requires rainy weather for the spores to germinate and penetrate, but rain, especially severe summer showers, also washes off the DVT. Within a matter of a day the levels can build up. But the one day when the DVT are washed off and the moisture is there is sufficient for the spores of the fungus to germinate and the fungus to penetrate and become established. Of course we weren't making our collections during the showers. We waited a few days and by that time the levels of DVT had recovered.

The moral of the story is that you have to do the field tests. You have to look at all aspects of the problem. With all the wonderful defences that plants and animals have, we still have troubles with pathogens.

Ley: This question of water is a very interesting one. It has always fascinated me because, for many of the antifeedants that we have worked with, one of the questions always is: if we spray onto a plant and there is a rain shower, do we lose material?

We have measured how quickly the antifeedants are absorbed into the leaf. It is also true that a shortage of water puts the plant under stress and natural antifeedants will build up during this time. Is there any evidence to suggest that all attack on a plant can be traced to water mobilization? For example, if you break or cut a leaf, the loss of water will be faster than normal. Does this trigger a protective biological response? When an insect feeds it could also impart chemicals into a plant, e.g. blood-sucking insects will stop the wound from healing so that there is a free-flow of blood supply. Is there a similar mechanism such that when insects attack a plant the cut does not seal rapidly and water continues to be lost?

Harborne: Pena-Cortes et al (1989) have suggested that the trigger molecule which causes all these changes in potato and tomato is abscisic acid. This is exactly the right plant hormone to be related to water since it controls stomatal opening. It is obviously not affecting stomata, it is acting as a messenger that's going through the plant, playing a different role to the one of controlling water loss.

Certainly water is very important. Plants which are normally under water stress, typically Mediterranean plants, tend to produce high concentrations of secondary compounds. This is very noticeable when you look at the essential oil content. We have identified all the herbs that grow in a certain Mediterranean area and determined which are rich in essential oils. Something like 38% are oil rich. In a temperate climate the number of plants rich in essential oils is much lower (11%). So there does seem to be an association between permanent water stress and production of essential oils.

Balick: In my observations of how indigenous peoples harvest plants that they are going to use for some sort of bioactivity, I notice that a significant per cent of the time they pull the first flush of growth, the young leaves. This is also carried out in the harvest of *Erythroxylum coca* for cocaine production and also commercially in the harvest of tea. Could we generalize and say that these people are capturing the maximum concentration of compound that they are seeking?

Harborne: It is dangerous to generalize about anything in this field but certainly that is noticeable. A herb is most vulnerable when it starts to grow, because at that time if it is eaten by anything it's going to disappear completely. Once it's over that first stage it is not as important—if it has twelve leaves and ten of them are eaten it can still survive. So plants are very vulnerable at that early stage and as you say there is evidence of high concentrations of active compounds in juvenile growth.

There is a close relationship between plants which seem to be well defended and medicinal plants.

Bowers: I was very interested in your statements about variable toxin defence. Did you mean that secondary chemicals may vary from structure to structure on the plant and this variability keeps the herbivore moving?

Harborne: Yes. This is the work of Schultz (1983) in the United States. I have only seen his reviews; I have never seen the original papers about it. He claims that if you measure the tannin content of the leaves along one branch of a beech or maple tree, you do find those sorts of variations.

Bowers: I really like that as an explanation for the variability. Many people measure something and cite the variability as a drawback and say that proves that these secondary compounds are really not important. If in a larger context the variability causes the herbivore to be exposed to increased predation, then it is a virtue for the plant.

Harborne: In some species of plants, particularly trees, you find that leaves are slightly damaged, which indicates that a herbivore has tried to eat a bit and decided against it. It is very difficult to see this in action a) because most insect herbivores eat at night, and b) generally speaking a leaf that is being eaten disappears within a matter of minutes; it is difficult to catch the intermediate stage.

influence and direct complex systems of natural interactions (Haslam 1988, Harborne 1988, Wink 1988).

Highly bioactive secondary plant products are involved in the various defence reactions that plants have developed against pathogenic or non-pathogenic microorganisms. Numerous constitutively produced secondary metabolites ('pre-infectional inhibitors'), such as various phenols, flavonoids, isoflavonoids, alkaloids, coumarins, cyanogenic glycosides or glucosinolates, afford the plant a certain degree of basic resistance. This is due to their pronounced antimicrobial activity. Furthermore, these compounds may accumulate to high concentrations in epidermal or cortical tissues of herbaceous plants (example: isoflavone malonylglucosides in chickpea, Fig. 3) and in bark or heartwood of trees (Harborne 1988, Wink 1988). Special attention has been given to the formation and rapid accumulation of low molecular weight, antimicrobial compounds ('phytoalexins') which are synthesized by plants *de novo* after infection by a microbial pathogen (Bailey & Mansfield 1982).

Phytoalexins are part of active plant defence mechanisms which are initiated in response to physical damage, various forms of stress or perception of a chemical signal derived from either the invading organism or the plant itself. Elucidation of the biochemical reactions involved in microbial attack of plants, in the spreading of infection and in the development of the induced plant defence responses will clarify the exact role of both partners in the interaction. Such investigations will also reveal the specific function of the newly synthesized phytoalexins. The differential accumulation of these antimicrobial compounds in compatible and incompatible plant–pathogen interactions plays a crucial role in the specificity of host resistance (Dixon 1986, Lamb et al 1989). Detailed knowledge of phytoalexins and their pattern of expression is not only of interest to the natural product chemist but will also contribute to modern breeding programmes to produce crop plants less prone to parasitic attack.

Elicitation of phytoalexins and other defence reactions

Plants have developed sophisticated active defence mechanisms against pathogens (Table 1). The main aims of these reactions appear to be inhibition of the microorganisms with antibiotic compounds and hydrolytic enzymes, inactivation of microbial exoenzymes with specific inhibitors and isolation of lesions by the deposition of lignin or by chemical modification of cell walls with polyphenols and/or hydroxyproline-rich glycoproteins. In most plant–pathogen interactions, these defence reactions are started simultaneously, though not all plants respond with the induction of the various routes with identical intensities. Tissue specificity in the expression of defence reactions is illustrated by comparisons of heterotrophic and photosynthetically active cells of *Nicotiana tabacum*—the former produce mainly the phytoalexin capsidiol, the latter make cell wall-bound phenolic compounds as their predominant response (Ikemeyer &

TABLE 1   Antimicrobial defence reactions of plants induced after infection

Chitinases and β-1,3-glucanases

Pathogenesis-related proteins

Inhibitors of fungal proteases, polygalacturonases and cellulases

Polyphenoloxidases and peroxidases

Phytoalexins

Hydroxyproline-rich glycoproteins and callose in cell walls

Isolation of lesions by lignin, suberin or phenolic polymers

Barz 1989). Numerous other changes in plant primary cell metabolism indicate that this is deeply affected by infection and totally re-directed to the expression of defence reactions (Dixon 1986, Lamb et al 1989, Threlfall & Whitehead 1988). Since most of the enzymes involved in the defence mechanisms (Table 1) are induced and synthesized *de novo*, the principle of differential gene activation is frequently an essential element of phytoalexin biosynthesis (Ebel 1986, Lamb et al 1989). Plant cell suspension cultures have been used with great benefit to investigate the early events of induction at the gene, mRNA or enzyme level (Hahlbrock & Scheel 1989) and to detect *de novo* synthesized products (Barz et al 1988).

Induction of phytoalexin formation in tissues or cell cultures results from the perception by the plant cells of elicitor molecules which may be of biotic or abiotic origin (Darvill & Albersheim 1984, Lamb et al 1989). Biotic 'exogenous' elicitors (polysaccharides, proteins, glycoproteins, unsaturated fatty acids) are structural components of pathogenic or non-pathogenic micro-organisms. Polysaccharide fungal elicitors are thought to interact with specific binding sites at plasma membranes, from which a signal is transduced to the nucleus (Ebel et al 1989). Oligomers of pectin from plant cell walls ('endogenous elicitors') may also induce defence reactions. These fragments are formed either by plant hydrolases or by microbial polygalacturonases. 'Abiotic elicitors' comprise a wide range of effectors (UV light, heavy metal ions, detergents, glutathione, xenobiochemicals, freezing or heating of plant cells), which exert various forms of stress and act in a mainly unknown way. Despite their pronounced toxicity for plant cells, heavy metal ions are excellent inducing agents for phytoalexins.

Structural aspects, function and biosynthesis of phytoalexins

There are nearly 300 known phytoalexins (Bailey & Mansfield 1982) and owing to continued efforts this number is rapidly increasing. A striking feature of phyto-alexins is their great structural diversity (Table 2); there seems to be no simple relationship between chemical structure and toxicity. The structural diversity is exemplified by the recently detected phytoalexins shown in Fig. 1. A particular

TABLE 2   **Selected examples of basic phytoalexin structures within the flowering plants**

Phenolic acid	Sesquiterpene	Naphthaldehyde
Biphenyl	Furanoterpene	Phenanthrene
Pyrone	Diterpene	Benzofuran
Chromone	Terpenequinone	Coumarin
Stilbene	Triterpene	Isocoumarin
Stilbene oligomer	Benzoxazin	Furanocoumarin
Flavonoid	Acridone	Polyacetylene
Isoflavonoid	Benzophenanthridine	Furanoacetylene
Pterocarpan		Thiophene

plant may produce a number of closely related compounds or the phytoalexins of one plant may be structural variants of compounds that accumulate constitutively (Fig. 2). Occasionally, chemically unrelated defence compounds have also been found in the same species (Harborne 1988). In general, in any given plant several substances are likely to be produced upon elicitation and a significant proportion of these structures are not otherwise known as natural products.

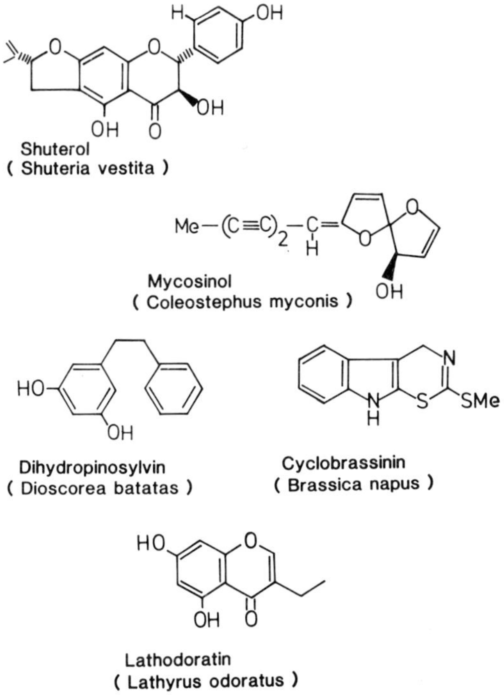

Shuterol
(Shuteria vestita)

Mycosinol
(Coleostephus myconis)

Dihydropinosylvin
(Dioscorea batatas)

Cyclobrassinin
(Brassica napus)

Lathodoratin
(Lathyrus odoratus)

FIG. 1.   Structural diversity of recently isolated phytoalexins.

At present, plant species of some 35 genera from at least 20 families have been analysed with respect to phytoalexin formation. There are approximately 400 families of flowering plants, therefore future research will undoubtedly reveal many new plant defence compounds. Apart from their biological function, this wealth of natural products will be of great interest for chemists, pharmacologists and plant biotechnologists. Phytoalexins are detected in plant extracts by fractionation procedures monitored by bioassays (inhibition of growth or viability of test microorganisms). Their inhibitory concentration, EC_{50}, was generally found in the range 10–500 µM, depending on the assay conditions.

The comparatively weak antimicrobial activity of phytoalexins and the highly varied concentrations observed in diseased plant materials have led to a controversial debate of the precise role of phytoalexins in plant defence reactions. Several lines of evidence now argue strongly for a causal role of phytoalexins in disease resistance. In general, phytoalexin accumulation proceeds more rapidly and to higher concentrations in incompatible interactions. Assays of the spatial distribution of phytoalexins at the infection site have demonstrated that the local concentrations may be sufficiently high to inhibit significantly hyphal growth (Hahn et al 1985). Screening different genotypes of one plant species, which show a varying degree of resistance to a particular fungal pathogen, for phytoalexin accumulation has repeatedly established a positive correlation between phytoalexin quantity and the extent of resistance. Inhibition of enzymes of phytoalexin biosynthetic pathways with molecular inhibitors turns an incompatible interaction into a compatible one (Moesta & Grisebach 1982).

FIG. 2. Structural comparison of constitutive components and induced phytoalexins isolated from *Erythrina crista-galli* (Mitscher et al 1988).

Suppressor molecules derived from fungal pathogens selectively prevent phyto-alexin accumulation (Barz et al 1989) and lead to induced susceptibility of otherwise resistant plant genotypes. Virulent pathogens appear to be more potent inducers of phytoalexins than are avirulent strains. In summary, although phytoalexin production may not be regarded as the main or only cause of plant resistance, such experiments convincingly indicate that these products represent an essential element in the total spectrum of defence reactions (Table 1).

Despite the well established antimicrobial activity of phytoalexins, it has not yet been satisfactorily explained why these compounds are toxic to fungi and even to mammalian and plant cells. There is a growing consensus that phytoalexin toxicity is a membrane-associated phenomenon and that the high lipophilicity of the phytoalexins is a structural prerequisite for their inhibitory potential (Laks & Pruner 1989). The various inhibitory reactions observed with phytoalexins all indicate that they interfere with normal functions of membranes (Giannini et al 1988). An important mechanism for the action of phytoalexins in eukaryotic cells is considered to be the uncoupling of mitochondrial oxidative phosphorylation (Laks & Pruner 1989).

Phytoalexin accumulation is a result of increased synthesis from remote precursors; the *de novo* formation of appropriate biosynthetic enzymes is characteristically observed. Intensive investigations of various stilbene, alkaloid, coumarin, isoflavonoid and terpenoid phytoalexins have provided substantial insight into the biosynthetic pathways, the enzymes involved and the molecular biology of enzyme induction (Dixon 1986, Ebel 1986, Hahlbrock & Scheel 1989, Lamb et al 1989). The complete biosynthetic routes of several phytoalexins have now been elucidated; one example is depicted in Fig. 3. In several cases, certain parts of phytoalexin biosynthesis proceed through routes which appear to be also required for the formation of constitutively accumulated compounds (see Fig. 3). Though formally identical, such parts of biosynthetic pathways are still subject to induction of enzyme activities by elicitors. A representative example is the biosynthesis of medicarpin and maackiain in *Cicer arietinum* (chickpea) (Fig. 3). In this pathway the coordinate induction of all enzymes from glucose-6-phosphate dehydrogenase through the stage of the formononetin intermediate to the terminal pterocarpan synthase has been observed (W. Barz et al, unpublished work 1989). Furthermore, in this plant five constitutive isoforms of chalcone synthase have been detected of which only two are substantially increased in their activities after elicitation of cell cultures (S. Daniel & W. Barz, unpublished work 1989). These and other results indicate that complex patterns of metabolic regulation are involved which are of special importance for enzymes at branch points of pathways. Convincing examples of a high level of poly-morphism of the gene systems coding for phytoalexin biosynthetic enzymes have been described (Hahlbrock & Scheel 1989, Lamb et al 1989).

The results depicted in Fig. 3 come from a programme to elucidate biochemical events and the essential plant resistance factors in the interaction between the

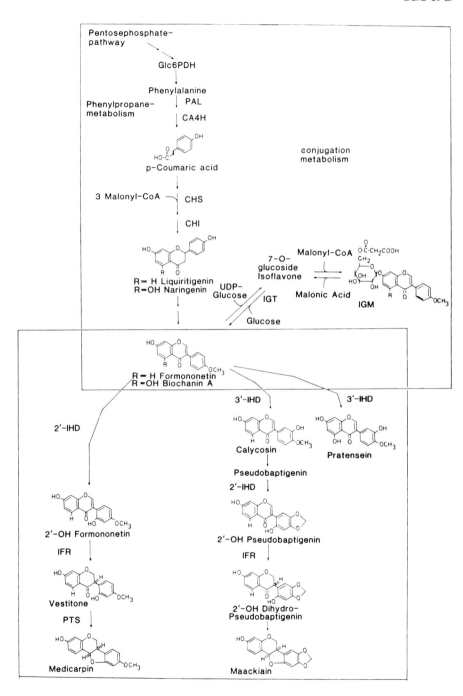

important crop plant, *C. arietinum* and its main fungal parasite, *Ascochyta rabiei*. Studies on *A. rabiei*-resistant (ILC 3279) and sensitive (ILC 1929) genotypes of chickpea revealed that several defence reactions are differentially expressed in the host. The phytoalexins medicarpin and maackiain consistently accumulate in much higher amounts in plants or cell cultures of the resistant cultivar (Barz et al 1988, 1989). Furthermore, β-1,3-glucanase and chitinase activities and other pathogenesis-related (PR) proteins (M_r 18 kDa) are preferentially induced in and secreted from cells of the cultivar ILC 3279. Upon elicitation of cell cultures a characteristic sequence of maximum activity of reactions (Lamb et al 1989) has been determined in which the levels of mRNAs for phytoalexin biosynthetic enzymes and for PR proteins peak after four hours, those of the enzyme activities and PR proteins after eight hours and the level of phytoalexin after approximately 12 hours (Barz et al 1989, S. Daniel & W. Barz, unpublished work 1989). In this temporal sequence, expression of β-1,3-glucanase and chitinase enzyme activities proceeds less rapidly; maximum levels are not reached until 24 hours after onset of elicitation (Vogelsang & Barz 1990).

An important step in phytoalexin biosynthesis is mediated by the isoflavone (formononetin) 2'- and 3'-hydroxylases (Fig. 3), which are the introductory enzymes of the pterocarpan-specific branch of biosynthesis. These enzymes are substantially induced in the cells of cultivar ILC 3279, whereas cells of the susceptible genotype show only very low hydroxylase activities. This striking difference has been interpreted as an essential limiting factor for the low phytoalexin production in cells of cultivar ILC 1929 (Barz et al 1989). Although biochanin A 2'- and 3'-hydroxylase activities have also been found to be elicitor induced, only the latter enzyme appears to act in the formation of the constitutive isoflavone, pratensein (Fig. 3). Other aspects of elicitor effects on constitutive isoflavone metabolism and its linkage with phytoalexin production in chickpea have been reviewed recently (Barz et al 1988, 1989, 1990).

Infection of plants or elicitation of cotyledons, seedlings or other tissues as well as cell cultures often affects the total secondary metabolism, leading to complex mixtures of newly synthesized compounds and increased levels of constitutive secondary products. In addition to phytoalexins themselves various products, such as intermediates of phytoalexin biosynthesis, compounds that

---

FIG. 3. (*Opposite*) Metabolic grid of constitutive phenylpropane metabolism and isoflavone conjugation reactions (upper box), and the inducible pterocarpan-specific pathway to the phytoalexins, medicarpin and maackiain (lower box) in *Cicer arietinum*. Upon elicitation all enzymes from glucose-6-phosphate dehydrogenase through the stage of the formononetin intermediate to the terminal pterocarpan synthase are concomitantly increased in specific activity by *de novo* synthesis with no effect on the enzymes of isoflavone conjugation. Glc6PDH, glucose-6-phosphate dehydrogenase; PAL, L-phenylalanine ammonia lyase; CA4H, cinnamic acid 4-hydroxylase; CHS, naringenin-chalcone synthase; CHI, chalcone isomerase; IGT, isoflavone-specific UDP-glucose:7-O-glucosyl transferase; IGM, isoflavone malonylglucosides; 2'-IHD, isoflavone 2'-hydroxylase; IFR, NADPH:isoflavone oxidoreductase; PTS, pterocarpan synthase.

normally accumulate in only trace amounts, and other biosynthetically unrelated substances, may accumulate in substantial amounts. This pronounced stimulatory effect of elicitation has been observed in investigations on isoflavonoids and their conjugates (Barz et al 1990), different alkaloids (Eilert 1987) or diverse monoterpenes and diterpenes (Croteau et al 1987). Such studies demonstrate that a clear distinction between elicitor-induced phytoalexins and elicitor-increased amounts of constitutive products is sometimes difficult to establish. However, such elicitation experiments represent a novel and very promising strategy to unveil the full potential of plants for secondary product formation. It is recommended that the analyses of elicitor-induced complex mixtures of secondary products by chemical, biological and pharmacological assays represent a technique for the detection of new bioactive plant products. This appears especially promising in cases of rare, tropical or otherwise difficult to obtain plant species. With regard to plant cell cultures, elicitation has successfully increased the yield of valuable bioactive products (Eilert 1987).

Phytoalexin detoxification by fungal pathogens

Phytoalexins are an active plant defence mechanism against invading microbes. Successful invasion of the plant host may in turn involve the ability of the pathogen to detoxify phytoalexins. All available data indicate that phytoalexin degradation is tightly linked to and determines one important trait of fungal pathogenicity (VanEtten et al 1989). Thus, phytoalexin detoxification may be important for the development of a particular plant disease.

The 3-O-demethylation of the pea phytoalexin, pisatin, by *Nectria haematococca*, (VanEtten et al 1989), the conversion of the bean phytoalexin kievitone to kievitone hydrate by *Fusarium solani* f. sp. *phaseoli* (Smith et al 1981), and oxidative and reductive conversions of the chickpea phytoalexins medicarpin and maackiain by *A. rabiei* (Höhl et al 1989) (Fig. 4) are examples from isoflavonoid phytoalexins. The established degradation of the potato sesquiterpenoid phytoalexins lubimin and rishitin by the potato pathogen *Gibberella pulicaris* (Desjardins et al 1989a), and the degradation by the same fungus of furanocoumarins (Desjardins et al 1989b) also demonstrate that tolerance of a pathogen to a phytoalexin is due to detoxification. Phytoalexin detoxification by fungal pathogens may result from either complete degradation (Höhl et al 1989) or only one or a few reaction steps (VanEtten et al 1989); the decisive feature, however, is that less toxic compounds are produced. The high degree of adaptation of a particular pathogen to the phytoalexins of its host is illustrated by the observation (R. Tenhaken & W. Barz, unpublished work 1989) that the chickpea pathogen *A. rabiei* will convert only the chickpea (6aR:11aR)-pterocarpans and not the (6aS:11aS)-isomers which also occur as natural products. Possible consequences of such specificity for the genetic engineering of more resistant plants have been discussed (VanEtten et al 1989).

FIG. 4. Initial degradative reactions of chickpea phytoalexins catalysed by pterocarpan:NADPH oxidoreductase and pterocarpan:FAD hydroxylase purified from *Ascochyta rabiei*. Both enzymes convert only (6aR:11aR)-medicarpin (R$_1$ = H, R$_2$ = OCH$_3$) and (6aR:11aR)-maackiain (R$_1$ = R$_2$ = O–CH$_2$–O) and not the (6aS:11aS)-isomers.

Knowledge of the enzymes of the initial detoxification reactions of phytoalexins in combination with modern molecular genetic techniques will lead to a better understanding of phytoalexin degradation for fungal pathogenicity (Leong & Holden 1989).

Pisatin 3-O-demethylation in *N. haematococca* is catalysed by a cytochrome P$_{450}$ monooxygenase. The gene for pisatin demethylase was cloned and expressed in the non-pathogen *Aspergillus nidulans*. The transformant, in contrast to the wild-type, actively demethylated pisatin but it was not able to infect pea plants (Schäfer et al 1989). However, expression of the gene for pisatin demethylase in the maize pathogen *Cochliobolus heterostrophus* conferred both the ability to degrade pisatin and a certain degree of pathogenicity towards pea plants (Schäfer et al 1989). Similar transformation experiments with this gene expressed in the chickpea pathogen *A. rabiei* (K.-M. Weltring et al, unpublished work 1989) also led to rapid pisatin demethylation, pronounced accumulation of the 6a-hydroxymaackiain product and a measurable virulence of the transformants in pea plants. Such properties are not shown by *A. rabiei* wild-type strains. These results provide direct evidence that in some host–parasite interactions phytoalexin degradation is essential for pathogenicity, although additional, unknown genetic elements are required for the complete determination of fungal virulence (Leong & Holden 1989). Furthermore, such investigations also elegantly prove that phytoalexins contribute to the resistance of plants towards microbial pathogens. Phytoalexins may be regarded as a reservoir of valuable bioactive compounds for practical application in agriculture or elsewhere. In this context, the degradation of these compounds by plant pathogens or saprophytes deserves special consideration. Detailed knowledge of the initial reactions of phytoalexin catabolism will contribute to the development of semisynthetic phytoalexins or analogous structures which are more resistant to biodegradation.

The detoxification reactions of pisatin by 3-O-demethylation and of the pterocarpans medicarpin and maackiain as shown in Fig. 4 are examples of this principle. Replacement of the methyl group in pisatin by an alkyl substitute (e.g. isopropyl) not readily removed in biological systems will render the molecule much less degradable. Furthermore, 6a:11a-dialkyl medicarpin or 3-O-isopropyl medicarpin will not be subject to either reductive conversion to 2'-hydroxyiso-flavans or oxidation to 1a-hydroxydienones. Application of this strategy may eventually lead to bioactive compounds more suitable for practical purposes.

Phytoalexin degradation and peroxidative polymerization

Accumulation of phytoalexins in infected tissues or elicitor-treated cell cultures is quite often a transient process followed by gradual disappearance of the compounds. This interesting observation requires explanation because at first sight one would expect phytoalexins to be persistent so that they could perform their antibiotic activity for longer.

Phytoalexin disappearance is tightly linked with the fact that these compounds are secreted from the producing cells into the extracellular space of tissues or, in case of cell cultures, into the growth medium (Barz et al 1990). Upon secretion phytoalexins are often confronted with remarkably high activities of extracellular acidic peroxidases (donor : H_2O_2-oxidoreductase, EC 1.11.1.7) (Barz et al 1990). In addition to the numerous catabolic reactions catalysed by peroxidases with constitutively formed secondary plant products, phytoalexins may also be subject to peroxidative degradation and/or polymerization (Barz et al 1990). If a substrate may form a phenoxy- or other radicals as an initial step, a peroxidase reaction proceeds smoothly as shown for several phytoalexins such as glyceollin, medicarpin or maackiain (Barz et al 1990). Peroxidase-catalysed polymerizations proceed either as an incorporation of the substrates into polymeric material (i.e. lignin, cellulose matrices) or as a self condensation process. Inspection of the chemical structures of phytoalexins leads to the assumption (Barz et al 1990) that the majority of the compounds are prone to peroxidative conversion. Thus, it is hypothesized that peroxidative processes are significant for the decrease of phytoalexin concentrations in the extracellular compartment of plant cells or tissues.

Additional support for such peroxidative processes is derived from the structures of various oligomers of the stilbene *trans*-resveratrol. These oligomers, which show different coupling orientation of the building moiety, are found as phytoalexins in grapevine ('viniferins') (Bailey & Mansfield 1982) or as bark constituents of *Stemonoporus* species (Bokel et al 1988). An additional feature of the peroxidase-catalysed polymerization reactions is that these processes involving radicals and hydrogen peroxide, may themselves exercise a strong antimicrobial activity that surpasses the antifungal toxicity of the genuine phytoalexins.

Conclusions

Research on phytoalexins and other plant defence reactions has provided significant insight into host–parasite interactions. As a major experimental tool, plant cell suspension cultures allowed investigations on the mode of gene activation, the expression of the phytoalexin response and the partial identification of the inducing molecules. Future investigations will undoubtedly lead to the discovery of new antimicrobial plant products which will be of interest for natural product chemists, pharmacologists and plant breeders. The modern techniques of molecular genetics will provide new strategies for the transformation of pathogenic fungi and crop plants. Genetic engineering programmes aiming at a more extensive expression of defence reactions in higher plants should focus on phytoalexins because of their importance among the determinants of plant resistance.

Acknowledgements

Financial support by Deutsche Forschungsgemeinschaft, Bundesminister für Forschung und Technologie, Bonn, Minister für Wissenschaft und Forschung, Düsseldorf, and Fonds der Chemischen Industrie is gratefully acknowledged.

References

Bailey JA, Mansfield JW (eds) 1982 Phytoalexins. Halsted Press, New York

Barz W, Daniel S, Hinderer W et al 1988 Elicitation and metabolism of phytoalexins in plant cell cultures. In: Applications of plant cell and tissue culture. Wiley, Chichester (Ciba Found Symp 137) p 178–198

Barz W, Bless W, Daniel S et al 1989 Elicitation and suppression of isoflavones and pterocarpan phytoalexins in chickpea (*Cicer arietinum* L) cell cultures. In: Kurz WGW (ed) Primary and secondary metabolism of plant cell cultures II. Springer-Verlag, Heidelberg, p 208–218

Barz W, Beimen A, Dräger B et al 1990 Turnover and storage of secondary products in cell cultures. Annu Proc Phytochem Soc Eur, in press

Bokel M, Diyasena MN, Gunatilaka AAL, Kraus W, Sotheeswaran S 1988 Canaliculatol, an antifungal resveratrol trimer from *Stemonoporous* [*sic*] *canaliculatus*. Phytochemistry (Oxf) 27:377–380

Croteau R, Gurkewitz S, Johnson MA, Fisk HJ 1987 Biochemistry of oleoresinosis. Plant Physiol (Bethesda) 85:1123–1128

Darvill AG, Albersheim P 1984 Phytoalexins and their elicitors. A defence against microbial infection in plants. Annu Rev Plant Physiol 35:243–276

Desjardins AE, Gardner HW, Plattner RD 1989a Detoxification of the potato phytoalexin lubimin by *Gibberella pulicaris*. Phytochemistry (Oxf) 28:431–437

Desjardins AE, Spencer GF, Plattner RD 1989b Tolerance and metabolism of furano-coumarins by the phytopathogenic fungus *Gibberella pulicaris* (*Fusarium sambucinum*). Phytochemistry (Oxf) 28:2963–2969

Dixon RA 1986 The phytoalexin response: elicitation, signalling and control of host gene expression. Biol Rev 61:239–291

Ebel J 1986 Phytoalexin synthesis: the biochemical analysis of the induction process. Annu Rev Phytopathol 24:235–264

Ebel J, Cosio EG, Grab H, Habereder H 1989 Stimulation of phytoalexin accumulation in fungus-infected roots and elicitor-treated cell cultures of soybean (*Glycine max* L). In: Kurz WGW (ed) Primary and secondary metabolism of plant cell cultures II. Springer-Verlag, Heidelberg, p 229–238

Eilert U 1987 Elicitation: methodology and aspects of application. In: Constabel F, Vasil JK (eds) Cell culture and somatic genetics of plants. Academic Press, San Diego New York vol 4:153–196

Giannini JL, Briskin DP, Holts JS, Paxton JD 1988 Inhibition of plasma membrane and tonoplast proton-transporting ATPase by glyceollin. Phytopathology 78:1000–1003

Hahlbrock K, Scheel D 1989 Physiology and molecular biology of phenylpropanoid metabolism. Annu Rev Plant Physiol Plant Mol Biol 40:347–369

Hahn MG, Bonhoff A, Grisebach H 1985 Quantitative localization of the phytoalexin glyceollin I in relation to fungal hyphae in soybean roots infected with *Phytophthora megasperma* f. sp. *glycinea*. Plant Physiol (Bethesda) 77:591–601

Harborne JB 1988 Introduction to ecological biochemistry, 3rd edn. Academic Press, London

Haslam E 1988 Secondary metabolism—fact and fiction. Nat Prod Rep 3:217–250

Höhl B, Arnemann M, Schwenen L et al 1989 Degradation of the pterocarpan phytoalexin (−)-maackiain by *Ascochyta rabiei*. Z Naturforsch Sect C Biosci 44:771–776

Ikemeyer D, Barz W 1989 Comparison of secondary product accumulation in phyto-autotrophic, photomixotrophic and heterotrophic *Nicotiana tabacum* cell suspension cultures. Plant Cell Rep 8:479–482

Laks PE, Pruner MS 1989 Flavonoid biocides: structure/activity relations of flavonoid phytoalexin analogues. Phytochemistry (Oxf) 28:87–91

Lamb CJ, Lawton MA, Dron M, Dixon RA 1989 Signals and transduction mechanisms for activation of plant defences against microbial attack. Cell 56:215–224

Leong SA, Holden DW 1989 Molecular genetic approaches to the study of fungal pathogenesis. Annu Rev Phytopathol 27:463–481

Mitscher LA, Gollapudi SR, Gerlach DC et al 1988 Erycristin, a new antimicrobial pterocarpan from *Erythrina crista-galli*. Phytochemistry (Oxf) 27:381–385

Moesta P, Grisebach H 1982 L-AOPP inhibits phytoalexin accumulation in soybean with concomitant loss of resistance against *Phytophthora megasperma* f.sp. *glycinea*. Physiol Plant Pathol 25:65–70

Schäfer W, Straney D, Ciuffetti L, VanEtten HD, Yoder OC 1989 One enzyme makes a fungal pathogen, but not a saprophyte, virulent on a new host plant. Science (Wash DC) 246:247–249

Smith DA, Harrer JM, Cleveland TE 1981 Simultaneous detoxification of phytoalexins by *Fusarium solani* f.sp. *phaseoli*. Phytopathology 71:1212–1215

Threlfall DR, Whitehead JM 1988 Co-ordinated inhibition of squalene synthetase and induction of enzymes of sesquiterpenoid phytoalexin biosynthesis in cultures of *Nicotiana tabacum*. Phytochemistry (Oxf) 8:2567–2580

VanEtten HD, Matthews DE, Matthews PS 1989 Phytoalexin detoxification: importance for pathogenicity and practical implications. Annu Rev Phytopathol 27:143–164

Vogelsang R, Barz W 1990 Elicitation of β-1,3-glucanase and chitinase activities in cell suspension cultures of *Ascochyta rabiei* resistant and susceptible cultivars of chickpea (*Cicer arietinum*). Z Naturforsch Sect C Biosci 45:233–239

Wink M 1988 Plant breeding: importance of plant secondary metabolites for protection against pathogens and herbivores. Theor Appl Genet 75:225–233

DISCUSSION

Cragg: I would like to relate a serendipitous discovery of a promising antitumour agent which resulted from a microbial–plant interaction. It was noted in Tennessee in the 1970s that cattle eating sweet potatoes infected with a *Fusarium* species of fungus were dying within a few days. They traced this to a compound that was acting on the lungs. Dr Michael Boyd isolated a very simple furan derivative which acts specifically on Clara lung cells. It is metabolized in lung cells to a highly electrophilic species which is the actual toxin. Ipomeanol is now in Phase I clinical trials as, hopefully, a selective agent against lung tumours. Ipomeanol is not formed by the plant or the fungus alone, but appears to be an elicitor-induced product.

Bellus: Professor Barz, a final objective of your research is to achieve expression of phytoalexins in plants. Is it not probable that pathogens will develop resistance to these compounds?

Barz: There is a little more behind our strategy. One aspect is, as you said, to transform plants so that they can produce other phytoalexins. This strategy has already successfully been used in Germany by Schröder & Schröder (1990). They used stilbene synthase, which is the key enzyme in the synthesis of the stilbene phytoalexin, resveratrol. They cloned the gene and expressed it in tobacco plants. The transgenic tobacco plants produce the stilbene phytoalexin, because the two precursors for this enzyme reaction, malonyl CoA and coumaroyl CoA are abundant in practically every plant.

I think further investigations on phytoalexins will show us what the relative importance of these compounds in the spectrum of the defence reactions actually is. If phytoalexins turn out to be very important, then perhaps we can use this strategy to make crop plants more resistant to their normal pathogens.

Secondly, we have seen that plants possess the ability to produce quite a number of defence compounds or to use different defence reactions. Phytoalexins are the first of these to be studied by molecular biologists because we know the biology and the enzymology quite well in comparison to those of some of the other defence mechanisms. Although, for instance, the work by Dr van Montagu and his colleagues on chitinases and β-1,3-glucanases has also progressed to the molecular biological and gene level.

Fowler: Wolfgang, you showed the induction of various enzymes of your biosynthetic pathway. Years ago, we worked on the very early enzymes of a related pathway in cell cultures. We were inducing the early stage of the shikimate pathway. What intrigued us was that we could get a very nice sequential induction of all those enzyme systems. At that time we hadn't the molecular biology available to look at the regulation of those enzymes. I wonder now if you have any thoughts about whether those enzymes are grouped together, are they co-regulated or what?

Barz: The enzymes of the pterocarpan biosynthetic sequence are co-ordinately expressed, not sequentially. We have studied a great number of other enzymes (Daniel et al 1988) in pathways of primary metabolism, such as glycolysis, citric acid cycle, oxidative respiration, and none of these enzymes is affected by elicitation. Therefore, it is a very specific effect on this particular biosynthetic pathway—except for glucose 6-phosphate dehydrogenase.

Fowler: This is a very important point because it could provide a good lead into switching on other secondary metabolite pathways. It may be the key we are looking for to express these systems.

Barz: I described the situation in the cell suspension culture where we do see accumulation of only phytoalexins with no other secondary metabolites. In the same situation in a tissue other secondary metabolites, not related to phytoalexins, are also increased in their concentrations. This is very nice strategy to unveil the full potential of a plant for secondary metabolite production.

Kuć: Could you say a little more about the pathogenicity? The ability of a mycelium to spread after insertion into a stem is not enough for pathogenicity.

Barz: The virulence test we used measures the spreading of the infection not the penetration of the fungus into the tissue, that's the great difference. For pathogenicity, other factors must be present. For instance, *Fusarium solari* and *Ascochyta rabiei* each produce an exocutinase. With the help of this cutinase, the hyphae can penetrate the tissue but other factors also are required. We have not produced a complete pea pathogen. We have only provided this *Ascochyta* fungus with a certain ability to spread in the pea plant and, we assume, to detoxify the phytoalexin.

Kuć: It was suggested some years ago that merely increasing the levels of phytoalexins in plants would be a way to protect plants against many diseases. This was approached in several ways. A search was made for plants that constitutively produced high levels of phytoalexins. It was also reasoned that elicitors of phytoalexin accumulation could be used. Some of these elicitors are fungal or plant cell wall oligosaccharides.

In all cases these approaches proved disastrous. The plants that constitutively accumulated phytoalexins—green beans, soya bean and pea—were inevitably stunted. The ones to which elicitors were applied that did produce high levels of phytoalexins were also stunted. The elicitors had to be applied extremely frequently because the phytoalexins are not translocated, the elicitors were not translocated and so all the new foliage had to be sprayed repeatedly. The phytoalexins are also degraded by fungi and by the plant itself. It is difficult to judge the reason for the stunting. One possibility is that there was diversion of energy and precursors to the synthesis of the phytoalexins and other processes therefore suffered. This doesn't mean that phytoalexins are not important; it doesn't mean that there are not ways in which we can use phytoalexins in plant defence. However, merely increasing phytoalexin levels has its hazards.

Barz: From our present point of view constitutive expression of a defence reaction is not a good strategy. What is essential is the ability of the plant to produce these compounds very rapidly, if necessary, and then at high concentrations.

Hall: It is certainly tempting to think that the expression throughout a plant of a new gene and its protein may be an energy drain. However, it is probably more the interference with the metabolism of the plant that is responsible for stunting the growth. This is probably why these phytoalexins are produced after infection. Dr Barz focused on their very high concentrations in the cell that is attacked.

In molecular biology we are moving from looking at single genes to whole gene sequences in intermediary metabolism, as has been elegantly described here, and to targeting. Gene expression is very specific to each cell. It is easy to think of just tissues but it's important to remember that each cell in a tissue has its own gene regulation. Clearly, some materials will pass from cell to cell and in specific ways. Sometimes even very small molecules are specifically prevented from moving across cell borders. The regulation of each cell is extremely important and has many implications with regard to the induction of disease resistance. These experiments show why we need more information on the specificity of gene expression and targeting.

In general many of these compounds appear to be advantageous, therefore why shouldn't a plant express many of these protections against heat stress, water stress, pathogen stress, all the time? It is clear that they interfere with the normal processes in the plant and I think that's probably the critical aspect.

McDonald: What do people in this field think about the potential of phytoalexins generally as leads for synthetic compounds? If I understood correctly, it's quite likely that these particular phytoalexins are phytotoxic and therefore the plant which produces them in marked quantities suffers a penalty.

Barz: The phytoalexins do have potential as lead structures for the synthesis of fungicides. All phytoalexins are somehow degraded by microorganisms. Detailed knowledge of these pathways is very helpful in devising the chemistry which we need for the biomimetic synthesis of fungicides. There is an example in the pterocarpans. The main microbial catabolic pathway of pterocarpans is either reductive cleavage of the dihydrofuran ring, leading to 2′-hydroxyisoflavans, or oxidation at position 1a. For either reaction to occur, there must be an hydroxyl group in position 3. If you block this hydroxyl group with an alkyl group, for example isopropyl which is not readily removed in biological systems, that gives a very potent fungicide with a comparatively high degree of resistance. An alternative is to substitute the pterocarpan in position

6a and/or 11a with an alkyl group, which will also block normal degradation and thus produce effective fungicides. I don't know how costly it would be to synthesize these compounds, but from a biological point of view they would be excellent fungicides.

References

Daniel S, Hinderer W, Barz W 1988 Elicitor-induced changes of enzyme activities related to isoflavone and pterocarpan accumulation in chickpea (*Cicer arietinum* L) cell suspension cultures. Z Naturforsch Sect C Biosci 43:536–544

Schröder J, Schröder G 1990 Stilbene and chalcone synthases: related enzymes with key functions in plant-specific pathways. Z Naturforsch Sect C Biosci 45:1–8

An economic and technical assessment of the use of plant cell cultures for natural product synthesis on an industrial scale

M. W. Fowler†, R. C. Cresswell° and A. M. Stafford°

†Department of Molecular Biology and Biotechnology, The University of Sheffield and °Plant Science Ltd, Firth Court, Western Bank, Sheffield S10 2TN, UK

Abstract. Plant cell cultures may be used as an alternative source of established natural products, as a source of novel 'lead' compounds or as a source of enzymes for modification of precursors. Only a few plant cell processes are operating commercially and their performance characteristics are industrial secrets. The economic aspects of natural product synthesis in plant cell cultures are presented on the basis of data derived from work on a pilot plant with bioreactors of 5–80 litres in which cells are grown in batch liquid culture. Cost analysis shows that the labour costs of operating plant cell culture processes are much higher than those for microbial processes, which reflects the longer process times of plant systems. These can be reduced by increasing the cell growth rate, the biomass yield and/or the product yield. Higher yields can be obtained by optimizing media conditions, but there are no standard guidelines for this. Each system has to be developed individually. Reducing the number of production runs a year, usually by increasing the number of days for which each batch of cells is synthesizing product, can markedly decrease costs. Economic assessment of the viability of production in plant cell cultures must consider not only production costs but also the expected market price of the product and the volume of sales.

1990 Bioactive compounds from plants. Wiley, Chichester (Ciba Foundation Symposium 154) p 157–174

In the past, plants have been a major source of pharmacologically active substances. Until the beginning of the twentieth century almost all medicinal agents were derived from plants, either as complex mixtures or as individually active components of variable degrees of purity. With developments in synthetic organic chemistry and, in more recent years, microbial biochemistry, the use of plants as a source of medicinal agents on an industrial scale has declined, although, as Farnsworth & Morris (1976) pointed out, in the USA at least 25% of prescription medicines are still derived from plants. A similar situation is also probably found in Western Europe.

The last decade has seen a renewal of interest in plants as a source of chemical structures with potential pharmacological activity. This resurgence of interest has occurred for a number of reasons, not the least of which is the increasing difficulty of creating novel structures through synthetic organic chemistry, coupled to a growing awareness of the potential of plants as a source of 'lead' chemical structures. A number of the major pharmaceutical companies have plant screening programmes. This commitment has been reinforced in many cases through concern about the rapid loss of what may be key germplasm during the clearing of the great tropical forests.

Another important factor is recent developments in plant cell technology. Significant progress has been made in plant cell culture over the last decade and a number (albeit a small one so far) of plant products are now produced in this way. Of these perhaps the best known is shikonin, a red pigment produced from *Lithospermum* cells by Mitsui and Co. (Fujita 1988a).

Plant cell cultures may contribute in three principal ways to the general area of natural product synthesis from plants:

1) as an alternative source of established products;
2) as a source of novel active principles ('lead' compounds);
3) as a source of biotransformation systems for the up-grading of synthetically or biologically derived precursor molecules.

In the situation where a novel structure has been identified and isolated from a cell culture, it should not be assumed that the cell culture will necessarily be the ultimate mode of production. Synthetic organic chemistry still features heavily in the production of 'natural products' and it is axiomatic that alternative approaches to product synthesis will be examined in the search for an economically viable process. The question then arises of the technical and economic viability of plant cell culture as a production route as compared with the principal alternatives of organic synthesis or derivation from whole plants. Technical considerations are placed first for the simple reason that without a technically viable process the economic considerations do not come into play. Equally, a technically feasible process may be developed which economically has no chance of competing with other processes.

Process overview

The overall approach to the establishment of plant cell cultures and process development has been well documented and will not be discussed further here (see e.g. Fowler 1986). In this paper we will focus on the process operation itself.

With only a few plant cell processes in commercial operation, there are few data available on which to carry out cost analysis. Also, those processes which are operational are maintained under great secrecy regarding their performance characteristics. Consequently, framework costings and economic assessments are hampered by a lack of detailed information. Many of the data used in this paper have therefore been derived from work on our own pilot plant with bioreactors of 5–80 litres. The results have been extrapolated to the sort of reactor volumes which might be anticipated for a 'typical' plant cell process operating in an industrial context. This should be borne in mind when considering the data and discussion which follow.

A technical or economic assessment of any process must take into account the overall time frame, as well as the process format. These will vary widely between product systems, and depend upon a variety of factors. For the purposes of illustration, we have chosen to focus on fairly simple systems developed in our own laboratory for the production of enzymes such as peroxidase and of secondary metabolites such as indole alkaloids from *Catharanthus roseus*. Fig. 1

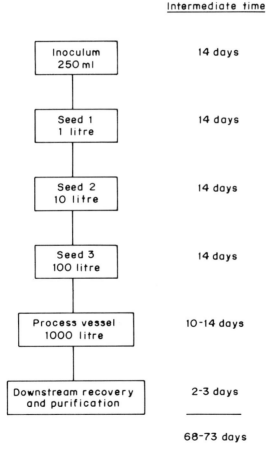

FIG. 1.   Process format for the production of a diagnostic enzyme by plant cell culture. The working volume at each stage is typically 70–80% of the nominal volume.

outlines the component parts of the process and indicates the time frame for each. A key point is that the cells are only in a 'useful' production phase for about 15% of the total time, but costs are of course incurred for the running of the whole process. This point has often been overlooked in the past when cost estimates for a plant cell culture process have been attempted. (Before proceeding, it is important to note that all the discussion in this paper will be based around plant cells grown in batch liquid culture. Immobilized cell technology would provide longer periods of continuous production.)

Economic aspects

An economic assessment in the early stages of a project is an important part of the overall programme, since it allows definition of the parameters within which one is working and also points to aspects of the process where economic advantage may be gained. This in turn serves two purposes. It identifies whether or not the proposed process has an economic advantage compared with existing production modes. Secondly, where a novel product is being assessed, which cannot be synthesized by other means, assessment shows whether or not it can be made at a price which the potential market will bear.

Cost calculations for production systems are complicated by both interrelated and independent variables, including cell biomass yields, percent product yields, number of bioreactor runs per year, bioreactor size, requirements for new equipment, and materials and maintenance.

The cost analysis presented has been based on a process being carried out in a 1000 litre batch production reactor with attendant seed vessels and cell harvesting and product recovery equipment. It has been assumed that all the required equipment would have to be purchased as new. As a rough guide, the cost of the equipment and its installation for a 1000 litre production system would be approximately US $726 000 (Table 1). Annual UK operating costs, including personnel, general maintenance, materials, power and depreciation (over 10 years), on the above capital cost are estimated to be approximately US $154 000 (Table 1).

Taking the above operating costs and assuming the following culture performance values:

1) 40×7 day contamination-free production runs per year,
2) a biomass yield of 12.0 g dry weight cells l^{-1},
3) a product yield of 1% of the biomass dry weight;

then from a 1000 litre vessel such a process would yield 4.8 kg of product per year at a cost of $32 170 kg^{-1}. Clearly this is a very high figure for many products. Attention was directed to those parts of the process where improvements and cost benefit potentials could be identified.

TABLE 1 **Estimated costs for a plant cell culture process**

Capital costs	US$1000	
Main bioreactor	500.0	
Inoculation bioreactor (100 litres)	117.0	
Services for installation, piping etc.	42.0	
Cell harvesting, disruption and extraction equipment	50.0	
Materials, shakers, glassware	17.0	
Subtotal	726.0	
Operating costs (per annum)		%
Personnel costs: Maintenance of reactor, preparation of cultures and media Extraction work	33.0	21.7
Materials, solvents for extraction (reusable)	8.0	5.4
Media components	6.0	4.0
Power costs: light, heat, pumps	17.0	10.9
Depreciation (@ 10%)	72.6	47.1
Repairs and maintenance contingency	17.0	10.9
Annual total	153.6	100.0

This large-scale process will require the presence of a smaller (100 litre) vessel to provide inocula for the larger reactor. Costings make provision for this. Cost assumes 40 production runs per year.

Table 1 shows that, in contrast to many microbial fermentation processes, the labour and associated costs of plant cell culture processes are very high, up to 50% of the total. This reflects the very long process times for plant cell culture operations. While it is difficult to reduce such operating costs directly, a number of possibilities exist. These may be considered under three headings:

1) cell growth rate,
2) biomass yield, and
3) product yield.

Analysis shows that improvements in biomass and product yield, as might be expected, alter production costs dramatically (Fig. 2). For example, biomass yields of $30 \, \text{g} \, \text{l}^{-1}$ with product yields of 2% reduce production costs to $6430 \, \text{kg}^{-1}$ (assuming, of course, that yields of 2% are readily achievable and maintained).

If fully depreciated equipment is used (i.e. no capital costs are incurred), then the cost of production at the same yield levels falls to $3410 \, \text{kg}^{-1}$. There is also economy in scale, and increased reactor size reduces costs accordingly.

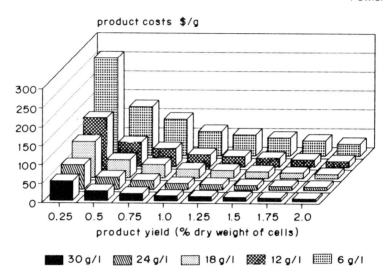

product costs $/g

product yield (% dry weight of cells)

■ 30 g/l   ▨ 24 g/l   ▧ 18 g/l   ▩ 12 g/l   ▤ 6 g/l

Assumes 40 runs per year at an annual
running cost of $155 K approximately

FIG. 2.   Effects of variations in biomass dry weight yields and percentage product yields
on production costs for plant cell culture process.

A smaller number of production runs per year with longer periods of 'productivity' also has a substantial effect on production costs (Table 2). These figures assume constant rates of cell growth and product synthesis for up to 16 days (which may not be true) and decreasing costs of personnel and some materials in proportion to the reduction in the number of runs. These costs, however, are also affected by changes in product yields. The calculated costs of production from a plant cell culture process using a range of yields and bioreactor runs per year are illustrated in Fig. 3. Calculations assume a modest biomass yield of $12.5 \, g \, l^{-1}$. The minimum cost of production by this system is $11 400 kg^{-1} using 20 runs per year with a product yield of 2% of the dry weight. Clearly, increases in the cell dry weight yield will reduce this price. Table 2 also indicates that annual estimated production levels rise with longer and fewer production runs, mainly because the number of non-productive days is reduced.

The assessment of the commercial viability of any plant cell culture process requires a comparison of the likely production costs with market prices for an established product, or a consideration of what the market will bear for a novel product. Many plant-derived drug components, for example the antineoplastic indole alkaloids such as vincristine and vinblastine, command market prices in excess of $5000 kg^{-1}. Other products are even more expensive. For instance the current market price for the pigment lycopene, from tomato, is over

TABLE 2 Effect of bioreactor run time on costings of production of compounds from plant cell culture systems

Runs per year	Maintenance days per year	No. days per run	Operating costs/yr US$1000	Costs per run US$1000	Final cell g/litre	g product per run	g product per year	Production costs $/g product
40	85	7.0	154	3.86	12.5	125	5000	30.87
35	75	8.3	148	4.24	14.8	148	5180	28.65
30	65	10.0	142	4.74	17.9	178	5355	26.58
25	55	12.4	136	5.45	22.2	221	5537	24.58
20	45	16.0	130	6.51	28.6	285	5710	22.81

Assumptions made are that 1) growth is constant at 1.78 g dry weight/day for at least 16 days; 2) product yield is 1% dry weight; 3) production vessel is 1000 litres. Two days per run are allowed for cleaning and refilling.

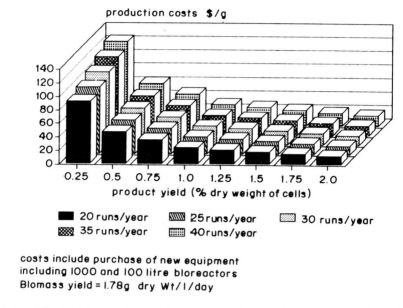

production costs $/g

20 runs/year   25 runs/year   30 runs/year
35 runs/year   40 runs/year

costs include purchase of new equipment
including 1000 and 100 litre bioreactors
Biomass yield = 1.78g dry Wt/l/day

FIG. 3.   Effect of the number of bioreactor runs/year and percentage product yields on production costs for a plant cell culture process.

$26 000 kg^{-1}. Such comparisons, however, must also consider market volumes (figures for which are often commercially sensitive), since it is of little value to have a highly priced product of which only a few kilograms are sold per year. A further consideration is that the ready availability of a product through a plant cell culture process, as compared with traditional sources, may initially depress market prices.

Technical considerations

The figures in Table 1 indicate that labour costs are a major component of plant cell culture process costs, which might prove difficult to reduce. However, as outlined above, there are technical approaches which indirectly may alleviate some of the problems of this situation. These will now be considered.

Growth rate

The overall time a process is operational and the period within this when a culture is 'productive' are a crucial part of the cost analysis for any process. This factor has long been accepted in microbial biotechnology but is only now beginning to be recognized in the area of plant cell culture (eg. Reinhard et al 1989). In the past, with plant cell culture rather too much attention has been focused on the period during which cells are actually synthesizing product rather than on the overall process time. A key goal must be to reduce the pre-production phase of cell growth to a minimum. There are a number of ways of approaching this, including the use of different process formats; however, the most fundamental is of increasing cell growth rate. Early work on plant cell culture was greatly hampered by the slow growth of plant cells. As recently as ten years ago, doubling times of three days and upwards were common. Numerous examples now exist of cell lines with cell cycle times of 24–48 hours, comparable to those of some fungal systems. Unfortunately, increasing growth rate typically results in lowered secondary product yield and so quite different approaches are often required to increase these two key culture performance parameters (Morris et al 1989).

Product yield and media manipulations

A characteristic of plant cell culture systems is that there are relatively few general rules for enhancement of product yields. Methods which are successful with one species (or sometimes one cell line) may not be applicable to others. Nevertheless, there are some parameters which, when altered, have raised yields in several systems (Table 3). For example, use of a high or optimum concentration of sucrose in culture media has enhanced yields of shikonin, rosmarinic acid and indole alkaloids from plant cell suspensions (Table 3). Similarly, replacement of 2,4-dichlorophenoxyacetic acid (2,4-D), which tends to inhibit secondary product formation, with other auxins such as 1-naphthalene acetic acid (NAA) or indole acetic acid (IAA), has been shown to produce higher yields of nicotine in *Nicotiana*, rosmarinic acid in *Anchusa* and indole alkaloids in *Catharanthus*. However, there are published exceptions reporting increased product yields after application of 2,4-D (Ikeda et al 1976) and decreased product yields in the presence of very high sucrose levels (e.g. 10%) (De-Eknamkul & Ellis 1985b).

Accumulation of some plant cell metabolites may be enhanced by addition of precursors, application of stress factors such as high salt concentrations or high light intensities, and depletion of macronutrients, such as nitrogen or phosphorus, in the medium. De-Eknamkul & Ellis (1985b) determined optimum concentrations of individual macronutrients required for the formation of rosmarinic acid by *Anchusa officinalis* cells. However, subsequent formulation of a medium incorporating all the pre-determined optimum concentrations of macronutrients resulted in only 60% of rosmarinic acid formation obtained in normal B5 growth medium. The authors concluded that intracellular conditions created when many nutritional factors are altered at the same time have different effects on metabolism than does alteration of a single component. Thus, determination of the most productive culture medium for a new plant cell culture process usually requires empirical experimentation based on experience of better

TABLE 3   Parameters for increasing yields

Product	Cells	Method of raising yield	Yield increase	Reference
Rosmarinic acid	*Anchusa officinalis*	Use of NAA as auxin	100% over use of 2,4-D	De-Eknamkul & Ellis 1985a
Rosmarinic acid	*Coleus blumei*	Use of optimum (7%) sucrose concentrations	87% increase compared with 2% sucrose	Zenk et al 1977
Rosmarinic acid	*Coleus blumei*	Addition of L-phenylalanine	100%	Ellis & Towers 1970
Rosmarinic acid	*Anchusa officinalis*	Alteration to macronutrients Sucrose (optimum 3%) Nitrate (optimum 15mM) Phosphate (optimum 3mM) use of NAA in place of 2,4-D	Yields varied from zero to 0.8 g/l depending on initial concentrations of macronutrients	De-Eknamkul & Ellis 1985b
Indole alkaloids	*Catharanthus roseus*	Addition of vanadyl sulphate	Sixfold increase, sometimes more (e.g. yields increased from 0 to 4.4 mg l^{-1})	Smith et al 1987
Indole alkaloids	*Catharanthus roseus*	Use of 8% sucrose and/or low concentrations of nitrogen	200% increase compared with 4% sucrose	Knobloch & Berlin 1980
		Replacement of 2,4-D with NAA	Serpentine yields: 0.1 (2,4-D) and 5.6 (NAA) mg/g dry weight cells	Morris 1986
Shikonin	*Lithospermum erythrorhizon*	High sucrose	100%	Mizukami et al 1977
Nicotine Anabasine	*Nicotiana tabacum*	Use of IAA as auxin	Synthesis inhibited by 2,4-D completely. Yields of up to 0.1% dry weight obtained with IAA	Furuya et al 1971
Serotonin	*Peganum harmala*	Feeding with tryptamine	16.5-fold increase	Sasse et al 1982

(continued)

TABLE 3 (*continued*)

Product	Cells	Method of raising yield	Yield increase	Reference
Cinnamoyl putrescines	*Nicotiana tabacum*	Depletion of phosphate	Production inversely correlated with phosphate concentration	Knobloch et al 1981
Caffeine	*Coffea arabica*	High light High NaCl	Up to 150-fold depending on culture type	Frischknecht & Baumann 1985
Ubiquinone	*Nicotiana tabacum*	Increase in 2,4-D from 0.05 to 2.0 mg l^{-1}	Approximate two-fold increase	Ikeda et al 1976
Valepotriates	*Valeriana wallichii*	Application of 0.05% colchicine	100-fold increase	Becker et al 1984
dopa	*Stizol-obium hassjoo*	Light application	Fivefold increase	Obata-Sasamoto & Komamine 1983
Berberine	*Thalictrum minus*	Addition of 6-benzyladenine (5–10 µM)	10-fold increase	Ikuta & Itokawa 1988

known systems. Successful determination of suitable conditions, however, can enhance yields considerably.

Altering the medium's composition of phytohormones is virtually guaranteed to affect secondary product formation. Similarly, the addition of product precursors nearly always has the expected positive effect on product yield. There have been spasmodic reports of yield enhancement in culture, achieved by imposing conditions that never would have been predicted to produce this effect. Amongst these are the use of vanadyl sulphate to enhance indole alkaloid accumulation in *C. roseus* cell cultures, and the use of colchicine to enhance the level of valepotriates in *Valeriana wallichii* cell suspensions (Table 3).

Culture stability and product yield

Plant cell suspension cultures generate somaclonal variation, thereby providing the source material for selection and improvement of cell line characteristics. This suggests that these culture systems are inherently phenotypically unstable. A genetic basis for this instability has been demonstrated in relatively few cases, although microscopic evidence for chromosomal aberrations is widespread.

The natural consequence of intraculture variability may be a temporal shift in the capacity of the culture to express secondary biosynthetic pathways.

Numerous examples have demonstrated that 'oscillation' around average productivities may occur with repeated subculture (Morris et al 1989) and that pressurized selection for particular characteristics may lead to the development of relatively unstable cell lines. This situation was found, for example, by Zenk in studies to select high indole alkaloid-producing cell lines of *C. roseus*. However, more recent work on protoplast-derived cell lines of *Lithospermum erythrorhizon* showed that the highly pigmented selected cell lines were more stable in their shikonin production than was the cell line from which they were derived (Fujita 1988b).

Cell line stability is an important consideration in economic terms from two standpoints. First, and most critically, variation in yield and in the timing of maximum product accumulation has a major impact on the final cost. In particular, the timing of final harvest for large-scale processing is critical in the case of an unstable product which may reach its maximum level during the growth phase. Secondly, the effort required to monitor and maintain selected cell lines cannot be underestimated. Present methods of plant cell line preservation generally depend on low temperature (e.g. $-195\,°C$) storage following a carefully monitored freezing schedule. Cell cultures of several species have been successfully stored under such conditions and in some cases their biosynthetic capabilities have remained unaffected, as in the case of *Anisodus acutangulus* (Zheng Guang-zhi et al 1983). In this example, lactalbumin hydrolysate and scopolamine yields before and after cryopreservation were very similar. Stability of biotransformation potential after low temperature storage has also been demonstrated for *Digitalis lanata* cell suspensions, which maintain their capacity to glycosylate β-methyldigitoxin to β-methyldigoxin (Diettrich et al 1982). However, as a standard procedure, low temperature plant cell line preservation does not offer the certainty of survival and retention of culture characteristics found in microbial or fungal cryopreservation. This means that to maintain targeted yields 'elite' cell lines must be treated with the greatest respect and maintained in their viable state by regular and consistent subculture.

While not directly related to cost projections for plant cell culture processes, instability in a cell line could have major implications in terms of operational costs and profitability. If it is felt that a cell line is inherently unstable, then it is wise to be conservative about yield estimates and to base cost estimates on moderately performing cell lines. Very high yields may then be regarded as a bonus.

Concluding remarks

An attempt has been made to provide a framework against which a general judgement and sensitivity analysis can be made for a plant cell culture process. Space has not allowed us to develop some of the aspects of this as we would have wished. We have attempted to pick out those factors which we believe show the best prospects for improving the economics of a plant cell process; however, there is no easy route and many other approaches (for instance, process format and cell immobilization) which may have something to offer in cost saving have not been covered.

What of the future? While there is no doubt that with an increased interest in plants as a source of 'lead' chemicals, a large number of novel structures with some degree of biological activity will be discovered, it is still questionable whether plant cell culture will be the major route of synthesis. Many compounds will lend themselves more readily to chemical synthesis; in other cases the plant cell system may be of greater importance in providing a key biotransformation enzyme. Having said this, the fact that we can discuss the use of plant cell culture as a possible and credible process option is an important step forward and of great significance to the pharmaceutical industry.

References

Becker H, Chavadej S, Baumer J, Stoeck M 1984 Isolation and characterisation of different cell lines of *Valeriana wallichii*. Proc 3rd Eur Congr Biotech, Munchen, vol 1:203–207

De-Eknamkul W, Ellis BE 1985a Effects of auxins and cytokinins on growth and rosmarinic acid formation in cell suspension cultures of *Anchusa officinalis*. Plant Cell Rep 4:50–53

De-Eknamkul W, Ellis BE 1985b Effects of macronutrients and rosmarinic acid formation in cell suspension cultures of *Anchusa officinalis*. Plant Cell Rep 4:46–49

Diettrich B, Popov AS, Pfeiffer B, Neumann D, Busenko R, Luckner M 1982 Cryopreservation of *Digitalis lanata* cell cultures. Planta Med 46:82–87

Ellis BE, Towers GHN 1970 Biogenesis of rosmarinic acid formation in *Mentha*. Biochem J 118:291

Farnsworth NR, Morris RW 1976 Higher plants; the sleeping giant of drug development. Am J Pharm 147:46–52

Fowler MW 1986 Industrial applications of plant cell culture. In: Yeoman M (ed) Plant cell culture technology. Blackwell, Oxford, p 207–227

Frischknecht PM, Baumann TW 1985 Stress induced formation of purine alkaloids in plant tissue culture of *Coffea arabica*. Phytochemistry (Oxf) 24:2255–2257

Fujita Y 1988a Industrial production of shikonin and berberine. In: Applications of plant cell and tissue culture. Wiley, Chichester (Ciba Found Symp 137) p 228–238

Fujita Y 1988b Shikonin: production by plant (*Lithospermum erythrorhizon*) cell cultures. In: Bajaj WPS (ed) Biotechnology in agriculture and forestry. Vol 4: Medicinal and aromatic plants I. Springer-Verlag, Berlin, p 225–236

Furuya T, Kojima H, Syono K 1971 Regulation of nicotine biosynthesis by auxins in tobacco callus tissues. Phytochemistry (Oxf) 10:1529–1532

Ikeda T, Matsumoto T, Noguchi M 1976 Effects of nutritional factors on the formation of ubiquinone by tobacco plant cells in suspension culture. Agric Biol Chem 40:1765–1770

Ikuta A, Itokawa H 1988 Berberine production through plant cell cultures. In: Bajaj WPS (ed) Biotechnology in agriculture and forestry. Vol 4: Medicinal and aromatic plants I. Springer-Verlag, Berlin, p 282–293

Knobloch K-H, Berlin J 1980 Influence of medium composition on the formation of secondary compounds in cell suspension cultures of *Catharanthus roseus* (L.) G. Don. Z Naturforsch 35c:551–556

Knobloch K-H, Beutnagel G, Berlin J 1981 Influence of accumulated phosphate on culture growth and formation of cinnamoyl putrescines in medium-induced cell suspension cultures of *Nicotiana tabacum*. Planta (Berl) 153:582–585

Mizukami H, Konoshima M, Tabata M 1977 Effect of nutritional factors on shikonin derivative formation in *Lithospermum erythrorhizon* callus cultures. Phytochemistry (Oxf) 16:1183–1186

Morris P 1986 Regulation of product synthesis in cell cultures of *Catharanthus roseus*. II. Comparison of production media. Planta Med 52:121–126

Morris P, Rudge K, Cresswell RC, Fowler MW 1989 Regulation of product synthesis in cell cultures of *Catharanthus roseus*. V. Long-term maintenance of cells on production medium. Plant Cell Tissue Organ Cult 17:79–90

Obata-Sasamoto H, Komamine A 1983 Effect of culture conditions on DOPA accumulation in a callus culture of *Stizolobium hassjoo*. Planta Med 49:120–123

Reinhard E, Kreis W, Barthlen U, Helmhold U 1989 Semicontinuous cultivation of *Digitalis lanata* cells: production of beta-methyldigoxin in a 300-L airlift bioreactor. Biotechnol Bioeng 34:502–508

Sasse F, Heckenberg U, Berlin J 1982 Accumulation of beta-carboline alkaloids and serotonin by cell cultures of *Peganum harmala*. Z Planzenphysiol 105:315–322

Smith JI, Smart NJ, Misawa M, Kurz WGW, Tallevi SG, DiCosmo F 1987 Increased accumulation of indole alkaloids by some cell lines of *Catharanthus roseus* in response to addition of vanadyl sulphate. Plant Cell Rep 6:142–145

Zenk MH, El-Shagi H, Ulbrich B 1977 Production of rosmarinic acid by cell suspension cultures of *Coleus blumei*. Naturwissenschaften 64:585–586

Zheng G-z, He J-b, Wang S-l 1983 Cryopreservation of calli and their suspension culture cells of *Anisodus acutangulus*. Acta Bot Sin 6:512–517

DISCUSSION

Farnsworth: More facts like this should be presented to bring scientists down to earth concerning the costs of research in the real world. I believe that if you are involved in drug discovery, you always have to be optimistic that there is somebody who is not in your area who can do something about your problems. Eli Lilly purchased leurocristine from Gedeon Richter (Budapest) in 1964 for $3.1 million/kg. They were selling it and making a profit. Now the Chinese are offering it at less than $3000/kg.

My bet is that your numbers are probably conservative.

Balick: You mentioned the difficulty of getting good material and how people send you things and it gets destroyed in the post. I do some work on the nuclear DNA of certain palms that are only found in the Amazon and for this we need fresh material. If you pack the leaves in whirl pack bags with one little sheet of tissue paper and send it via DHL from virtually any country in the world it can be in the lab in three days at the most. You always have good material and it is a remarkable delivery service.

Morinda citrifolia has been grown in culture; what is it used for?

Fowler: There is no commercial use for this plant in Europe; I don't know about in Japan.

Balick: So something is being produced from that plant in culture but you say there's no use for it. Why is it in a bioreactor?

Fowler: A number of those systems have been worked on to try to develop a commercially viable cell line. In a number of cases there have been dramatic increases in the yield of product but they are still insufficient to put into commercial production.

Yamada: At the Nippon Paint Company, Dr Y. Yamamoto is culturing *Euphorbia millii* tissue (Yamamoto et al 1989). He has found one very pure anthocyanine (cyanidine monoarabinoside) which has never been found previously in tissue culture. If we extract anthocyanine from flower petals, we get a mixture of anthocyanines, but using his system he could produce a single pure cyanidine monoarabinoside which has a beautiful colour.

I am very impressed by your production of peroxidase in cell culture. Do your cultured cells accumulate the peroxidase within the cell or secrete it into the medium?

Fowler: When we screened our culture bank for peroxidase, we found that some cell lines from different species release the peroxidase into the medium while others retain it in the cells. However, we found that we could manipulate the degree of secretion. The ratio of enzyme activity between cells and medium also changed with age. The cell line that we have releases about 85% of the total peroxidase at the point of maximal synthesis into the medium. So it's a useful commercial cell line in that particular aspect.

Yamada: Peroxidase has many isozymes; how do you isolate the particular one you want?

Fowler: That particular cell line produces two major isozymes of peroxidase. The key one is usually termed the neutral peroxidase. It is thermostable, has quite a flat pH profile and is very useful in diagnostics.

Yamada: The price of different peroxidase isozymes is quite different. The neutral one is quite expensive, so you are onto a good thing!

van Montagu: I agree that with plant cell cultures it's wise to go for the products made in cells you can culture, and not hope for a product that is made in specialized cells within the intact plant. Because those cells will de-differentiate in culture.

So what do you do if you want something made in a specialized cell type? Chemical synthesis can be prohibitively expensive or even impossible if there are too many asymmetric carbon atoms. I see quite a future for genetic engineering in plants. Professor Barz has shown that some pathways are co-regulated. If you concentrate on the last enzymes, you can isolate the gene quite easily. You do not have to purify the enzymes, you just have to enrich the fractions. From only a few micrograms on a 2D gel of proteins you can isolate enough material to do amino acid sequencing. Then you can synthesize a nucleic acid probe and go for the gene. When you have the genomic sequence, you take the promoter and fuse it to two different marker genes (e.g. hygromycin phosphotransferase and β-glucuronidase) and introduce both constructs (on the same T-DNA) into plant cells. By screening for mutant transformants that overexpress both marker genes, it should be possible to isolate mutations in the corresponding *trans*-acting factors rather than mutations in the promoter *cis* sequences. We are working only on the experimental stages of this but it should soon be possible.

Fowler: In Professor Barz's lecture I was most interested in the coordinate expression of those enzyme systems. If we can begin to see the expression of whole enzyme sequences, and find one event which will switch on those pathways, then we may have ways of making in bacterial systems traditional plant products normally produced in plantations.

Yamada: Tissue culture will still be important. For example, peroxidase is a haem-containing peptide. It is also a glycoprotein and the sugar side chain is necessary for its activity. Therefore, even once you have the gene, it will be very difficult to produce this enzyme in bacteria. You will still need plant cell cultures.

Fowler: I agree; the question I thought we were addressing is whether or not we could find a promoter system to switch on a whole enzyme pathway to make a non-protein product—an alkaloid or a steroid. I doubt that we would see the cloning of maybe 10 genes for a whole pathway into a microbial system. It would be more profitable to go for a specific promoter for a coordinator gene to switch on the pathway inside the plant cell for product synthesis.

van Montagu: It is being found more and more that upstream promoters consist of a series of sequences to which *trans*-acting proteins bind. These transcription factors can have either a stimulatory or an inhibitory effect on gene expression. Some have been identified and it's clear that they act on the regulatory sequences of co-regulated genes, such as genes encoding enzymes in the same synthetic pathway. So in future we will not try to express these pathways in bacteria, we want to enhance their expression in the plants.

We have also done a lot of work on peroxidases. The many peroxidase isozymes are targeted to different positions within the cell and some are secreted. Proteins are made in the endoplasmic reticulum and secreted unless they possess signals to direct them to a specific compartment within the cell. We have identified six peroxidase isozymes in tobacco. Three are acidic and are automatically secreted; the others are more basic and have signals to target them to vacuoles, where they accumulate. If you fuse the sequence of an appropriate signal peptide with the coding sequence of any foreign protein (e.g. neomycin phosphotransferase) and you introduce this construct under the control of suitable regulating signals into plant cells, then the foreign protein can be secreted (Denecke et al 1990). It's fascinating how many enzymes you can get secreted from plant cells in culture.

Fowler: It would be interesting if we could isolate those secretory signals and attach them to other enzyme gene sequences. We could then produce a range of enzymes which would be excreted into the medium—that would make the biochemical engineering job of recovery much more simple.

van Montagu: That's what we have done and it's really easy. Unless there is an additional sequence in your protein that would target it to an an intracellular compartment (nucleus, peroxisome, vacuole, etc.), the enzyme will be secreted. A possible problem is the presence of sequences which become glycosylated during transport. For example, glucuronidase is secreted but inactivated because it has an asparagine residue necessary for enzyme activity which becomes glycosylated.

Potrykus: I cannot see how any fermentor can be cheaper than plants in the field, even with the proposal of using genetic engineering to identify the key enzymes for specific pathways. I assume that also with this strategy it would be cheaper to grow the transgenic plants in the field and harvest the material from the plants. Plants are optimized for using sunlight, not for being grown on sucrose.

I can see some advantages with systems where the product can be excreted but I am afraid that the calculation presented by Professor Fowler is far too optimistic. It is based on the exceptional model system of a shear-resistant plant cell line, which dates back 25 years roughly, and this does not produce many interesting compounds. In cases where you have to establish a new culture, you have to invest a few more years and these cultures probably will not be shear resistant.

It would be interesting to see realistic calculations of the costs of harvesting 1kg of precious material from plants growing in the field or in the forest, and 1kg extracted from plant cells in a fermentor.

Fowler: I agree with you in that I don't think it's sensible to go looking at the moment for established plant products. I agree that to go back to the field is the correct approach. The case I would make for cell cultures is to return to enzymology and to say that if one wanted a small supply of a key enzyme, this may be a better way of producing that small supply on a continuous basis. We need to be very focused in the application of cell culture technology. In the past claims for its potential have been too widespread and perhaps exaggerated. Whether, as Dr van Montagu says, we can identify these switching sequences and promoter sequences to get whole enzyme pathways operational, time will tell. At the moment the figures I have quoted are probably conservative, and the costs are still far higher than the costs of traditionally produced plant products. I would argue that plant cell culture is not yet viable for the production of many currently available products.

van Montagu: The only example we have at the moment of making a peptide at what seems to be a commercially competitive rate is the synthesis of [Leu]-enkephalin peptide as part of a storage protein (Vandekerckhove et al 1989).

The 2S storage proteins constitute 30% of seed weight. It is possible to substitute a stretch of amino acids in this 2S protein and still obtain correct synthesis, folding, deposition in protein bodies and stability in storage. In a pilot experiment, seven amino acids in the 'variable region' of the 2S protein were substituted by the [Leu]enkephalin peptide. This peptide was flanked by two lysine residues to facilitate recovery (by trypsin and carboxypeptidase B digestion) of the peptide from the extracted storage protein.

From our results in the lab, we estimate that one hectare of engineered rape seed will yield the present total global production of [Leu]enkephalin. Recently, we deleted 23 amino acids in the storage protein and substituted them with a new sequence of 36 amino acids. Hence, this approach looks promising and could allow production of a whole set of biologically important peptides.

References

Denecke J, Botterman J, Deblaere R 1989 Protein secretion in plant cells can occur via a default pathway. Plant Cell 2:51–59

Vandekerckhove J, Van Damme J, Van Lijsebettens M et al 1989 Enkephalins produced in transgenic plants using modified 2S seed storage proteins. Bio/Technology 7:929–932

Yamamoto Y, Kinoshita Y, Watanabe S, Yamada Y 1989 Anthocyanin production in suspension cultures of high-producing cells of *Euphorbia millii*. Agric Biol Chem 53:417–423

General discussion II

Chemical synthesis of bioactive compounds from plants used in traditional Chinese medicine

Liu: Traditional Chinese medicine and folk medicine in China has a history of more than 2000 years. Recently, more emphasis has been laid on research on bioactive products from the plants used medicinally. We have a joint programme to study 200 plants commonly used in traditional Chinese medicine.

We studied several species of the *Glycyrrhiza* genus. More than half of the prescriptions in traditional Chinese medicine include extracts from this genus. We isolated several flavonoids and some triterpenoids. It is very unusual in that the plant contains flavonoids of different skeleton types, such as isoflavonoids and dihydroflavonoids, with different combinations of sugar moieties, including monoglycosides, diglycosides, *O*-glycosides and *C*-glycosides. It also contains some chalcones and medicarpin.

We also studied some species of genus *Epimedium* and isolated a series of flavonoids with a prenyl side chain at the 8 position and different combinations of sugar moieties, including some new compounds. The compound, heperoside, is a commonly encountered flavonol galactoside. We isolated this compound in very high yields from some species of *Rhododendron*. We tested its biological activity and clinical tests show that is an effective antitussive drug. We have a project to synthesize this compound from quercetin, but so far the yield is low.

We worked on some species of *Scutellaria*, another commonly used herb in Chinese traditional medicine. We isolated a series of flavonoids and dihydroflavonoids with a very unusual B-ring oxygenated substitution pattern.

Some species of the genus *Swertia* have been used for industrial production of preparations to treat acute viral hepatitis. We have worked on some species of this genus and isolated a series of xanthone glycosides, including two novel xanthone dimers. This is the first discovery of natural xanthone dimers with a C-C linkage.

There are some new drugs which we developed during the last two decades. One is anisodamine. The chemical structure is quite similar to that of hyoscyamine, but anisodamine has an extra hydroxyl group at position 6. It is very interesting because it has lower toxicity than atropine. It has been produced commercially and is available on the market. The synthetic product has more side effects than the natural one.

Another drug is anisodine, which has a similar structure to scopolamine, but with an additional hydroxy group on the side chain. Its pharmacology is very interesting. Animal tests and clinical observations have shown that both anisodamine and anisodine improve the microcirculation.

The third new drug we developed during the last decade is a compound isolated from *Fagopyrum cymosum*. It is the only ingredient included in a secret prescription owned by a herbalist. We isolated the main active principle, flavon-3-ol dimers, in very high yield, about 7%. This preparation has been produced industrially for treatment of lung abscess. It was used to treat a case of lung cancer in 1977, in a patient for whom radiotherapy and chemotherapy were no use, because the disease had progressed too far. The patient's wife was a friend of my colleague and asked if we had any drug that could help. This compound has very low toxicity so you can give large doses. The patient was given 2 g three times a day. After three weeks the patient was so improved that the doctor didn't believe it was the same patient! After two months this patient was fully recovered.

We discussed with some doctors in the cancer hospital whether we could introduce this compound for the treatment of lung cancer. It is very depressing because the cancer patients cannot be treated with a drug that has not been proved in Western-style trials. So we still have problems to see what we can do with this compound.

So some drugs used in folk medicine are very effective, and it is important to prove their efficacy and see if they work in Western medical trials.

Opportunities for bioactive compounds in transgenic plants

Timothy C. Hall, Mauricio M. Bustos, Janice L. Anthony, Li Jun Yang†, Claire Domoney† and Roderick Casey†

Biology Department, Texas A&M University, College Station, TX 77843-3258, USA, and †Applied Genetics Department, John Innes Institute, Colney Lane, Norwich NR4 7UH, UK

Abstract. A variety of bioactive compounds have now been introduced into plants through recombinant DNA techniques. Early examples included genes encoding proteins conferring herbicide tolerance and insect or virus resistance. More recently, pharmacologically useful compounds such as enkephalin and immunoglobulin have been produced in transgenic plants. Modification of existing compounds to provide better nutritional value or improved functional properties is exemplified in the case of seed storage proteins. The value of RNAs as bioactive compounds for suppression of undesirable products and viral infection has now been demonstrated in plants. The developmentally regulated expression of novel bioactive compounds in defined tissues, and their targeting to specific subcellular locations, is becoming of ever increasing economic and sociological importance as knowledge of the molecular mechanisms involved accumulates.

1990 Bioactive compounds from plants. Wiley, Chichester (Ciba Foundation Symposium 154) p 177–197

Progress in the development of techniques for gene transfer to higher plants has been rapid over the past decade. Approaches based on *Agrobacterium tumefaciens* (Ti plasmid) have moved from conceptual methodologies to reliable, routine, procedures used widely by industrial and university research groups and even in undergraduate level laboratory exercises. The first generation of commercial crop plants modified by recombinant DNA technologies will soon be available. The range of crop plants transformed is impressive. Nevertheless, much remains to be learned concerning extension of this range. Procedures for direct introduction of DNA by electroporation (Potrykus 1988, Davey et al 1989), microprojectile bombardment (Klein et al 1988) and microinjection (Neuhaus et al 1987) are being actively investigated (Potrykus, this volume). Such procedures are potentially applicable to monocot crops, few of which appear susceptible to transformation by conventional approaches using the Ti plasmid.

A major reason for the introduction of novel genetic information into plants of economic value is, of course, its expression as bioactive compounds conferring new or enhanced traits. In this review, we show that RNAs as well as proteins are being used as bioactive macromolecules. Uses range from novel approaches for inducing resistance to herbicides, insects and various pathogens, to enhancement of nutritional value and development of wider tolerance to environmental conditions. As will be explained, greater understanding of the determinants conferring tissue, temporal and developmental specificity of gene expression is needed. Similarly, further knowledge concerning the factors which contribute to the levels of desirable gene products and their targeting to functional sites is essential. The identification of new genes and biological processes capable of producing bioactive compounds will be an ongoing aspect of plant genetic engineering.

Compounds active in crop protection

Herbicides

The potential of genetically engineering plants to be resistant to various herbicides was recognized some years ago (Comai et al 1985, Shah et al 1986). Induction of resistance to herbicidal chemicals often provides the opportunity to identify and select successfully modified progeny cells, since non-transformed tissues are killed by the herbicide or antibiotic.

Two different approaches for plant protection against herbicides have been exploited. Detoxification, in which genetic information encoding a bioactive compound capable of degrading the herbicide is introduced into the target plant, has proven to be very effective. Degradation of the herbicidal antibiotics kanamycin and G418 by neomycin phosphotransferase (Bevan et al 1983, Fraley et al 1983) remains the most widely used example of the detoxification approach. Introduction of nitrilase encoded by a gene isolated from *Klebsiella ozaenae* (Stalker et al 1988) leads to effective degradation of bromoxynil, a commercially used herbicide; tests using field-grown tobacco plants have confirmed that there is no yield penalty. Indeed, enhancement of yield under normal farming conditions can be expected through effective weed control provided by this herbicide. Induction of herbicide resistance has also been accomplished via the detoxification approach for the phosphinothricin herbicide, Basta™, by expression of the phosphinothricin acetyltransferase gene from *Streptomyces hygroscopicus* (De Block et al 1987). Recently, detoxification of 2,4-dichlorophenoxyacetic acid by introduction of a gene from *Alcaligenes eutrophus* encoding 2,4-dichlorophenol 6-monooxygenase (EC 1.14.13.20) into tobacco was reported (Lyon et al 1989), suggesting that the use of auxin-based herbicides can be extended to genetically engineered broad-leafed crops.

A second approach for protection against herbicides is the introduction of a bioactive compound that is unaffected by the herbicide, and capable of substituting functionally for the endogenous target compound. Glyphosate herbicides such as Round-Up™ are very effective because they interfere with the shikimic acid biosynthetic pathway that is essential in plants. Glyphosate binds and inactivates the native plant enzyme, 5-enolpyruvylshikimate-3-phosphate (EPSP) synthase, which is intrinsic to this pathway. Comai et al (1983) found that the bacterial gene aro-A encodes an enzyme that is relatively unaffected by glyphosate. Introduction of aro-A (Comai et al 1985) or mutated forms of plant EPSP synthase (Shah et al 1986) has yielded transgenic plants resistant to application of glyphosate. However, the undegraded herbicide accumulates in growing apices and yield penalties have been encountered using the currently available genes. Resistance to the sulphonylurea herbicides, chlorsulfuron and sulfometuron methyl, has been accomplished by expression of acetolactate synthase from *Arabidopsis thaliana* (Haughn et al 1988).

Insecticides

Insect infestations cause major losses in crop production. Loss occurs directly through destruction or impairment of the growing plant or plant product (e.g. the seed) by herbivorous insects and their larvae, and indirectly when insects introduce fungal, bacterial or viral pathogens. A parasporal protein toxic to lepidopteran insects is copiously produced by *Bacillus thuringiensis*, and insecticides effective against more than 50 lepidopteran pest species (Wilcox et al 1986) incorporating spore or crystal protein preparations have been sold commercially for many years. Several research groups have transferred DNA encoding this protein to higher plants. The level of bioactive compound accumulated in transformed plants is typically rather low, and sequence modification of the toxin mRNA (perhaps also of the protein) may be necessary to improve stability. Nevertheless, Delannay et al (1989) found that, in field tests, transgenic tomato plants expressing the toxin from *B. thuringiensis* var. *Kurstaki* were substantially protected against tobacco hornworm (*Manduca sexta*); significant control of tomato fruitworm (*Heliothis zea*) and tomato pinworm (*Keiferia lycopersicella*) was also observed.

Additional compounds reported to be bioactive against insects include lectins from beans (*Phaseolus vulgaris*). Arcelin, first described by Romero (1984), is strongly implicated as the bioactive agent protecting certain cultivars of bean seeds against bruchid beetles (*Zabrotes subfasciatus*) during storage (Osborn et al 1988). Moreno & Chrispeels (1989) have reported that the *P. vulgaris* αAI inhibitor of insect α-amylases is encoded by a gene very similar or identical to a lectin gene characterized by Hoffman et al (1982). These proteins are considered to be lectins on the basis of their high sequence homology with lectins whose carbohydrate-binding properties have been well characterized.

Broekaert et al (1989) have reported the isolation of a chitin-binding lectin from stinging nettle (*Urtica dioica*) that is antagonistic to several fungi pathogenic to higher plants. Brown et al (1982a,b) have detailed erythrocyte-agglutinating, leucocyte-agglutinating and mitogenic activities of lectins from several *P. vulgaris* cultivars. The sequence similarity of these bioactive lectin-like proteins suggests that purification of a specific peptide from a protein mixture by physical separation procedures may be impossible. Expression and biological testing of the protein encoded by cDNA or genomic sequences may be required for rigorous confirmation that a given biological activity is due to the isolated clone. Careful sequencing of the coding regions of these clones is intrinsic to such characterization, and caution is advisable in assessing whether bioactive properties described in existing publications on lectin and lectin-like activities are correctly assigned to a given protein or DNA sequence where such characterizations have not been undertaken.

The use of trypsin inhibitors for protection of crop plants against insects has been suggested (Newmark 1987). Protection of cowpea (*Vigna unguiculata*) seeds against bruchid beetles (*Callosobruchus maculatus*) has been attributed to the presence in certain lines of a Bowman-Birk trypsin inhibitor (Gatehouse et al 1979, Hilder et al 1989), and it appears that several classes of protease inhibitors are potentially useful bioactive compounds for plant protection (Garcia-Olmedo et al 1987).

Pathogen antagonists

Chemical protection is widely used to reduce crop losses from fungal and bacterial infections. While control of these pathogens by bioactive compounds introduced by genetic engineering is likely in the future, current excitement concerns recombinant DNA approaches for control of viruses. There is no effective chemical control for viruses, and it is difficult to estimate their effect on crop yield except in extreme cases, such as rice tungro where yield reduction can be over 70% (Ling 1972). Another virus disease, Hoja blanca, caused widespread loss of the rice crop in Venezuela in 1956 and has caused total loss of a year's crop in Cuba (Grist 1986). Discovery of this disease on rice in Florida in 1957 resulted in a total ban on rice cultivation in that state for several years.

The first definitive example of protection of a transgenic plant against virus infection by a bioactive compound was provided by Powell-Abel et al (1986) who showed that expression of tobacco mosaic virus coat protein could delay disease development in transgenic tobacco plants. Viral coat protein had previously been implicated in the phenomenon of cross-protection, where infection of a given plant by a mild virus strain prevents subsequent infection by a severe strain of the same virus. The protective value of expressing viral coat protein in transgenic plants has now been demonstrated for a wide range of virus groups. These include alfalfa mosaic virus (Tumer et al 1987,

Loesch-Fries et al 1987, Van Dun et al 1987), cucumber mosaic virus (Cuozzo et al 1988), potato virus X (Hemenway et al 1988, Hoekema et al 1989), and the tobraviruses, tobacco rattle and pea early browning (Van Dun & Bol 1988). Stark & Beachy (1989) recently reported protection against soybean mosaic virus, a member of the potyvirus group, in transgenic tobacco expressing soybean mosaic virus coat protein. This finding is significant because many potyviruses are pathogens of economically important plants and because it implies that broad, multivalent resistance can be induced by the expression of viral coat proteins in plants.

Systemic infection by viruses is dependent upon the expression in the infected plant of a movement or transfer (*tra*) protein. The function of such *tra* proteins has been related to their interaction with plasmodesmata in alfalfa mosaic virus (Stussi-Garaud et al 1987) and tobacco mosaic virus (Deom et al 1987, Wolf et al 1989). Expression of modified *tra* proteins in transgenic plants may provide a way to limit systemic spread of viruses.

A radically different approach for protection against virus infection is provided by the expression of satellite RNAs in transgenic plants. Satellite RNAs may have little apparent sequence similarity to the carrier virus upon which they are dependent for their replication, but their presence can dramatically increase or decrease the severity of pathogenesis. Expression of cucumber mosaic virus (CMV) satellite in transgenic tobacco provided substantial protection against infection by CMV (Baulcombe et al 1986); Gerlach et al (1987) showed that expression of tobacco ringspot virus satellite protected transformed tobacco plants against the carrier virus. This approach is attractive because the amount of protective satellite synthesized increases by RNA replication pathways once an infection occurs, whereas the amount of coat protein available for cross-protection depends on the strength of the eukaryotic promoter used in the construct. Concerns about the approach include the possibility that mutations of the beneficial satellite sequence will result in accentuation rather than amelioration of symptoms. As with many transgenic systems, refinements to this approach can be expected in the near future. However, the efficacy of the systems tested so far confirms that RNAs can function as beneficial bioactive compounds in transgenic plants, and additional examples are given below.

Environmental protection

Protection of crop plants against adverse environmental conditions is a complex challenge; nevertheless, examples exist that permit optimism in this area. The discovery that certain strains of *Pseudomonas syringae* and other bacteria are very effective ice nucleation agents (Lindow et al 1978) suggested novel ways for plant protection against frost injury. The gene(s) encoding the nucleating proteins have been modified by recombinant techniques. It is hoped that spraying bacteria containing these genes onto plants will induce extracellular (as opposed to intracellular) freezing and provide a degree of protection. Enhanced chilling protection may be achieved by targeting expression of genes encoding these proteins to the epidermis of transgenic plants by recombinant DNA and plant transformation techniques.

Functional modification of seed proteins

A combination of classical plant breeding studies and electrophoretic character-ization of wheat seed proteins has provided much insight into molecular parameters involved in bread-making qualities (reviewed by Payne 1987). Two high molecular weight glutenin subunits (numbers 10 and 12) were found to confer different degrees of elasticity to dough (Tatham et al 1985, 1987). These have since been cloned (subunit 12: Halford et al 1987; subunit 10: Flavell et al 1989). Comparison of the amino acid sequences encoded by these genes shows that subunit 10 has a more regular β-turn structure over a longer region, producing a stronger β spiral with better elastic properties. Flavell et al (1989) note that this information may be used to predict changes to the subunit 10 sequence that will enhance its contribution to dough quality. Introduction of suitably modified genes into wheat is a challenge for the future. Dough made from sorghum seed flour is almost totally deficient in visco-elastic properties essential for the leavening process. It would be interesting to determine if artificial mixtures of subunit 10 protein and available sorghum flour yield better dough. If so, then introduction and expression of genes encoding subunit 10 (or improved variants) in sorghum may dramatically increase the potential of this crop.

Cereal seed proteins are typically low in lysine whereas legume proteins have low methionine content. An attempt was made to change this situation for the bean (*P. vulgaris*) storage protein, phaseolin, by the insertion of a sequence corresponding to that from a 15 kDa zein (a storage protein from *Zea mays* seed) that encodes 15 amino acids, six of which are methionine (Hoffman et al 1988). Although the amount of mRNA expressed in transgenic tobacco seed was similar to that predicted, only low levels of protein accumulated. This suggests that the modified protein structure is susceptible to proteolysis, supporting the theory that alteration of a given storage protein to enhance its nutritional value will best be accomplished when based on a knowledge of the tertiary structure of the native protein (Blagrove et al 1984, Lawrence et al 1990). Successful enhancement of the methionine content of transgenic tobacco seed has been achieved by expression (using the phaseolin promoter) of a brazil nut seed protein coding sequence that contains 18% methionine residues (Altenbach et al 1989).

Novel uses of plants for the production of biologically active compounds

The feasibility of producing biologically active pharmaceuticals in transgenic plants has been established. Vandekerckhove et al (1989) replaced part of an

Arabidopsis thaliana 2S seed storage albumin gene with sequences encoding the neuropeptide [Leu]enkephalin, bounded by tryptic cleavage sites. Up to 200 nmol of peptide were recovered per gram of seed from *A. thaliana* and *B. napus* plants transformed with this construct.

Sexual crosses between transgenic tobacco plants expressing either γ or κ immunoglobulin chains yielded progeny in which both chains were simultaneously expressed and assembled into functional antibody that accumulated to levels of 1.3% of total leaf protein (Hiatt et al 1989). Low levels of protein were detected when constructions lacking the mouse leader sequences were used for transformation, suggesting that the endoplasmic reticulum or Golgi systems in plant cells can recognize immunoglobulin chains. The catalytic IgG$_1$ antibody (6D4) selected binds a low molecular weight phosphonate ester, so functionality of the assembled antibody could be tested. The authors noted that accumulation of functioning antibodies in transgenic plants could provide an array of new options, including the introduction of new catalytic activities into existing metabolic pathways.

Novel uses for proteases

Viral proteases

The nutritional role of proteases in complete digestion of protein substrates to amino acids is well known. However, the physiological regulation of many important enzyme functions by limited proteolysis is less well recognized. Zymogens are converted to active compounds by partial hydrolysis: examples include many digestive enzymes (e.g. trypsinogen, chymotrypsinogen) that respond to a foreign stimulus, hormones for which processing is a normal physiological response (e.g. proinsulin) and proteins that are cleaved in accordance with developmental programming (such as procollagenase, procollagen and plasminogen). Thus, proteolytic processing is an effective post-translational mechanism that can rapidly respond to a signal to change the activity and destination of a target protein (Neurath 1989).

An intrinsic part of many viral life cycles is the cleavage of a precursor 'polyprotein' into several proteins that have independent activities. This was first recognized for animal viruses, as was the fact that the protease(s) involved was itself encoded by the virus and lay within the polyprotein (Palmenberg et al 1979). Many plant and animal viruses are now known to encode proteases and to require polyprotein processing in their life cycle (Kräusslich & Wimmer 1988). Although many of the proteases involved in viral polyprotein processing are encoded by the virus, host proteases may also be involved (Carrington et al 1990). Interference with virus-specific proteolytic processing systems is an attractive new target for plant biotechnology. Several avenues towards this goal appear feasible. For example, expression of peptides containing the specific processing site would result in competition with viral polyproteins for the viral protease; a more effective approach might be to design the site so as to permit binding, but not release, of the viral protease. Arrest of translation of the viral protease region by expression and hybridization of its antisense RNA might also be effective.

Seed protein proteases

Legumins, seed storage proteins found widely in the seed of legumes, crucifers and even in oats, are synthesized as a protein whose endoproteolytic cleavage yields acidic (α) and basic (β) subunits that undergo disulphide bonding to yield α/β dimers (Fig. 1A). Because legumins are often encoded in multigene families, it has been difficult to assign specific peptides to individual genes and consequently to assess which genes are major contributors to protein stored in the seed.

Results of a novel experimental approach to this problem for pea legumin are shown in Fig. 2; this provides a strategy for accumulation of high levels of bioactive compounds (Fig. 1B). Studies *in vivo* (Chrispeels et al 1982, Croy et al 1982, Domoney & Casey 1984) have shown that pea (*Pisum sativum*) legumin is synthesized as a precursor of M_r 80 000 (Fig. 2 i) or 60 000 (Fig. 2 ii); in the majority of food legumes the M_r 60 000 precursor is most abundant. Cleavage by a cellular endoprotease yields peptides of sizes shown diagrammatically in Fig. 2 ii and iii. Translation *in vitro* of mRNAs hybrid-selected by pea legumin clones pCD32, pCD43 and pCD40 (Domoney et al 1986,

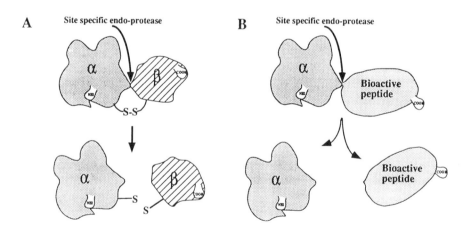

FIG. 1. Diagrammatic representation of polyprotein processing. Panel A shows the legumin precursor which yields an acidic (α) and a basic (β) subunit after site-specific cleavage by an endoprotease. The disulphide (S–S) bond is easily broken in the presence of reducing agents. Panel B shows a similar situation in which the β subunit is replaced by a bioactive peptide.

Domoney & Casey 1987) yielded polypeptides of M_r 80 000, 60 000 and 63–65 000, respectively (Fig. 2a,c and the first dimension gel in Fig. 2e). After treatment *in vitro* with a protease obtained from immature pea cotyledons (C. Domoney, unpublished), the M_r 80 000 precursor yielded predominantly polypeptides of approximately M_r 25 000, 23 000 and 20 000 (Fig. 2b); the sizes of two of these subunits correlate with those derived from 'small' legumin dimers of M_r 35 000 (Matta et al 1981). After processing, the M_r 60 000 precursor yielded polypeptides of approximately M_r 40 000 and 20 000 (Fig. 2d), indicating that the products of the pCD43 class constitute the majority of the legumin. Considerable homology exists between pCD32 and pCD40 and the products selected by pCD40 also contain some M_r 80 000 precursors (first dimension gel in Fig. 2e). The M_r 63–65 000 precursors yielded a heterogeneous group of polypeptides with M_r of approximately 40 000 and 18–20 000 (substantial amounts of precursor are unprocessed). Hybrid-selection can be used to isolate a class of mRNAs, but it cannot give an absolute correlation between a given polypeptide and a specific DNA sequence. For this, *in vitro* transcription and translation of a defined clone is required: an example is shown in Fig. 2f, where the pea legumin clone p10 yields a discrete polypeptide of M_r 60 000. Treatment of the *in vitro*-generated polypeptide with the pea protease preparation gives products that include the expected M_r 40 000 α subunit.

The transcription–translation approach described above permits investigation of the precise nature of, and constraints on, the cleavage site; site-directed changes can be introduced into the p10 DNA clone and polypeptides with modified amino acid sequence(s) at the processing site can be obtained. Suitable design of the sequence near the proteolytic cleavage site should allow release of the β subunit (or a substituted bioactive peptide such as the enkephalin sequence mentioned previously) from the α subunit. Release of the bioactive compound would simplify its subsequent purification. Targeting of the seed storage protein to protein bodies in the developing seed is important for sequestration and stability. It is possible that information for this targeting is contained within the α subunit; thus a bioactive peptide substituted for the β subunit might be stably sequestered in high concentration within the protein bodies of the seed.

RNAs as bioactive molecules

The use of viral satellite RNAs to modulate pathogenesis has been discussed above. A wider application of bioactive RNAs in plant biotechnology being explored by many research groups (reviewed by van der Krol et al 1988b) is the expression of antisense RNAs—RNA molecules of opposite polarity to metabolically active RNAs. Advantages offered by this system include a high specificity for selection of the target sequence; disadvantages include the apparent requirement for a large excess of antisense sequence. The feasibility of this

Published observations from in vivo data

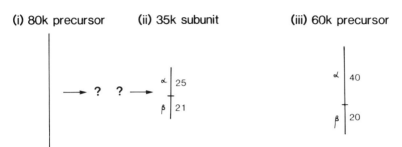

(i) 80k precursor   (ii) 35k subunit   (iii) 60k precursor

Observations by hybrid-selection/translation/processing

pCD32 class   pCD43 class   pCD40 class

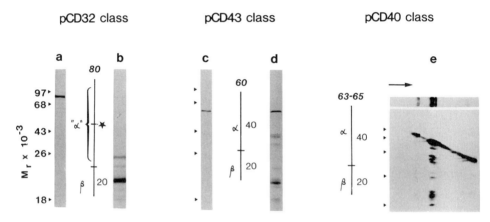

Observations by in vitro transcription/translation/processing

f   g

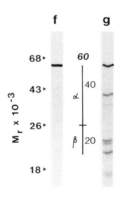

FIG. 2. *Caption on opposite page.*

approach has been demonstrated by the reduction of colour in petunia flowers by expression of antisense mRNA encoding chalcone synthase (van der Krol 1988a). The antisense approach is also being explored as a means to reduce softening of tomato fruit. Polygalacturonase activity has been implicated in this process and partial success in reducing polygalacturonase levels has been reported by Sheehy et al (1988) and by Smith et al (1988) through the expression of antisense mRNA in transgenic tomatoes. These reports indicate that antisense RNAs can function as bioactive molecules in plants, and that their activities may be very useful in reducing, perhaps eliminating, undesirable proteins in crop plants.

Examples of the use of antisense RNAs to induce protection against viruses include the expression of complementary RNA sequences from genomes of cucumber mosaic virus (Rezaian et al 1988), potato virus X (Cuozzo et al 1988, Hemenway et al 1988) and tobacco mosaic virus (Powell et al 1989) in transgenic tobacco. Unfortunately, these have proven to be of limited value, giving protection only when low levels of inoculum are supplied. Further modifications of the RNA sequences used and the ways in which they are expressed may give better results. In order to combat establishment of invading pathogens, catalytic rather than stoichiometric activities of the antisense sequence are desirable, and this has stimulated adaptation of the catalytic self-splicing (ribozyme) functions of introns (Cech 1987), viroids (Forster et al 1988), virusoids (Forster & Symons 1987) and viral satellites (e.g. tobacco ringspot satellite: Buzayan et al 1986). A general procedure for deletion of undesirable genes (including viral genomes) has been developed that incorporates ribozyme function. This is known as the 'molecular shears' concept, in which the substrate and catalytic functions of tobacco ringspot satellite RNA are separated, the catalytic sequence being flanked at both sides by sequences complementary to the target RNA (Haseloff & Gerlach 1988). The only constraint appears to be that the target (substrate) sequence must contain -GUC-, the conserved sequence adjacent to the cleavage site.

The use of mutant RNA sequences to investigate replication strategies of RNA viruses revealed that, *in vitro*, some mutants showed higher ($-$) strand promoter activities than those of wild-type RNAs. However, *in vivo*, these mutants were less infectious than the wild-type controls (Bujarski et al 1986). Recent results in our laboratory suggest that the decrease in infectivity may result from tight

---

FIG. 2. (*Opposite*) Experimental evidence for proteolytic processing of pea legumin precursors. Data obtained from observations *in vivo* are shown on the top line (i–iii); lanes a–e show the complexity of legumin peptides resulting from hybrid-selection, translation and processing; lanes f and g show the application of transcription, translation and processing to elucidate the origin of various peptides and the mechanism of processing. Vertical lines represent precursor polypeptides; proposed cleavage sites between the α and β subunits are indicated by short horizontal lines. Numbers indicate $M_r \times 10^{-3}$. Further details are given in the text.

binding of the mutant sequence to host enzymes, preventing release and recycling of the viral RNA. Such experiments provide a basis for optimism that the expression of certain mutant viral RNAs in transgenic plants may provide a novel, potent, replication interference strategy for crop protection.

Targeting, stability and regulation of bioactive molecules

The concept of tissue (and cell) specificity of expression was recognized before the development of recombinant DNA methodology. Nevertheless, cloning and *in situ* hybridization techniques have accentuated the fact that gene expression is absolutely cell specific. This is not so evident in tissues where many cells are expressing the same product, but is clear in the case of petunia EPSP synthase which is abundantly expressed only in floral tissues (Benfey & Chua 1989). An important consideration raised by such findings is that protection of a plant against herbicide action may require that *every* cell normally affected adversely by the herbicidal compound be protected. Indeed, it is possible that protection will be best achieved when expression is in the correct *subcellular* location. This may be a specific membrane surface within the cell, or within an organelle such as the chloroplast. Expression must also occur under a wide range of environmental conditions, so a promoter that works well only in bright sunlight or at moderate temperatures may fail to confer adequate protection under field conditions.

The above considerations demand increased understanding of the determinants for transcriptional expression of a given gene, and of the stability and targeting of its mRNA and protein. Regions extending 500 bp or more upstream of the coding sequence of several plant and animal genes contain sequence motifs that act in various ways to modulate gene expression (Benfey & Chua 1989, Bustos et al 1989). Like the TATA element that lies approximately 30 bp upstream of the transcriptional start site of most eukaryotic genes, these elements may be very short. Proteins and other *trans*-acting factors that participate in signal transduction from a stimulus to gene expression bind to these sequences and to each other in ways that are only just beginning to be understood. Tabata et al (1989), Katagiri et al (1989) and Hartings et al (1989) have cloned several such *trans*-acting proteins from plants. It will be important to learn if these proteins are unique for a given type of expression, or if permutation of common *trans*-acting factors can yield different effects. The latter seems most likely, otherwise the number of regulatory factors would have to equal the number of tissue-specific genes. It is tempting to think that interference with *trans*-acting factors, or introduction of many DNA-binding domains that would deplete the cell pool of a given factor, may produce major differences in gene expression.

Despite our imperfect knowledge about determinants for gene expression, it is clear that transfer of coding regions to transgenic plants under the control of certain DNA sequences can result in protein expression in novel sites. Thus, placement of the coding region from a seed-specific β-conglycinin gene under the control of a 5′ flanking sequence from the cauliflower mosaic virus 35S gene resulted in expression in leaves and other non-seed tissues (Lawton et al 1987). However, little protein accumulated in the leaves. Such experiments accentuate the need for targeting gene products to specific sites as one element in stability; even when transcription and translation of seed proteins occurs effectively, lack of sequestration of the protein product may result in its rapid degradation.

Information regarding targeting signals on proteins is becoming available. This includes targeting of proteins to plant mitochondria (Boutry et al 1987) and chloroplasts (Karlin-Neumann & Tobin 1986); other studies have defined sequences that direct nuclear targeting (Dingwall et al 1987). Many proteins bear complex carbohydrate (glycan) moieties, and it is tempting to believe that these residues provide important targeting signals (Paulson 1989). However, in the bean (*P. vulgaris*), glycoprotein phytohaemagglutinin alteration by site-directed mutagenesis of one or both of the canonical N-glycosylation sites prevented glycosylation but did not interfere with targeting of the proteins to protein bodies in transgenic tobacco seed (Voelker et al 1989). These experiments suggest that the vacuolar targeting signal in phytohaemagglutinin lies in the peptide sequence. Although the hydrophobic leader sequence of some 20 amino acids found at the N-terminus of many eukaryotic proteins that are sequestered into membrane-bound structures has been well characterized, no explicit sequence specificity is apparent.

Summary

This account provides a brief description of the range of bioactive compounds already expressed in transgenic plants. Recombinant DNA engineering is providing novel opportunities for increasing our basic knowledge of biological form and function and for exploiting crop plants that exhibit important properties. Indeed, new ways to generate bioactive compounds in plants are described frequently in this fertile area for innovative minds. Nevertheless, much remains to be understood in regard to the control and processing of bioactive proteins and RNAs and the regulatory signals leading to their functional expression.

Acknowledgements

Aspects of these studies were supported by a collaborative NATO grant (RG85/0742) to R.C. and T.C.H. Funding from NSF (Grants DCB 8802057 and DCB 8904886), Texas Advanced Technology Program (Grant 9999 02202), Rhône-Poulenc Agrochimie and the Texas A&M Office of University Research is gratefully acknowledged.

References

Altenbach SB, Pearson KW, Meeker G, Staraci LC, Sun SM 1989 Enhancement of the methionine content of seed proteins by the expression of a chimeric gene encoding a methionine-rich protein in transgenic plants. Plant Mol Biol 13:513–522

Baulcombe DC, Sanders GR, Bevan MW, Mayo MA, Harrison BD 1986 Expression of biologically active viral satellite from the nuclear genome of transformed plants. Nature (Lond) 321:446–449

Benfey PN, Chua N-H 1989 Regulated genes in transgenic plants. Science (Wash DC) 244:174–181

Bevan M, Flavell RB, Chilton M-D 1983 A chimeric antibiotic resistance gene as a selective marker for plant cell transformation. Nature (Lond) 304:184–187

Blagrove RJ, Lilley GG, Von Donkelaar A, Sun SM, Hall TC 1984 Structural studies of a french bean storage protein: phaseolin. Int J Biol Macromol 6:137–141

Boutry M, Nagy F, Poulsen C, Aoyagi K, Chua N-H 1987 Targeting of bacterial chloramphenicol acetyltransferase to mitochondria in transgenic plants. Nature (Lond) 328:340–342

Broekaert WF, Van Parijs J, Leyns F, Joos H, Peumans WJ 1989 A chitin-binding lectin from stinging nettle rhizomes with antifungal properties. Science (Wash DC) 245:1100–1102

Brown JWS, Osborn TC, Bliss FA, Hall TC 1982a Bean lectins. Part 1: relationships between agglutinating activity and electrophoretic variation in the lectin-containing G2/albumin seed proteins of French bean (*Phaseolus vulgaris* L.). Theor Appl Genet 62:263–271

Brown JWS, Osborn TC, Bliss FA, Hall TC 1982b Bean lectins. Part 2: relationship between qualitative lectin variation in *Phaseolus vulgaris* L. and previous observations on purified bean lectins. Theor Appl Genet 62:361–367

Bustos MM, Guiltinan MJ, Jordano J, Begum D, Kalkan FA, Hall TC 1989 Regulation of β-glucuronidase expression in transgenic tobacco plants by an A/T-rich, *cis*-acting sequence found upstream of a French bean β-phaseolin gene. Plant Cell 1:839–853

Buzayan JM, Gerlach WL, Bruening G 1986 Satellite ringspot virus RNA: a subset of the RNA sequence is sufficient for autolytic processing. Proc Natl Acad Sci USA 83:8859–8862

Carrington JC, Freed DD, Oh C-S 1990 Expression of potyviral polyproteins in transgenic plants reveals three proteolytic activities required for complete processing. EMBO (Eur Mol Biol Organ) J 9:1347–1353

Cech TR 1987 The chemistry of self-splicing RNA and RNA enzymes. Science (Wash DC) 236:1532–1539

Chrispeels MJ, Higgins TJV, Spencer D 1982 Assembly of storage protein oligomers in the endoplasmic reticulum and processing of the polypeptides in the protein bodies of developing pea cotyledons. J Cell Biol 93:306–313

Comai L, Sen L, Stalker D 1983 An altered aroA gene product confers resistance to the herbicide glyphosate. Science (Wash DC) 221:370–371

Comai L, Facciotti D, Hiatt WR, Thompson G, Rose RE, Stalker DM 1985 Expression in plants of a mutant *aro*A gene from *Salmonella typhimurum* confers tolerance to glyphosate. Nature (Lond) 317:741–744

Croy RRD, Lycett GW, Gatehouse JA, Yarwood JN, Boulter D 1982 Cloning and analysis of cDNAs encoding plant storage protein precursors. Nature (Lond) 295:76–79

Cuozzo M, O'Connell KM, Kaniewski W, Fang R-X, Chua NH, Tumer NE 1988 Viral protection in transgenic tobacco plants expressing the cucumber mosaic virus coat protein or its antisense RNA. Bio/Technology 6:549–557

Davey MR, Rech EL, Mulligan BJ 1989 Direct DNA transfer to plant cells. Plant Mol Biol 13:273–285

De Block M, Botterman J, Vandeweile M et al 1987 Engineering herbicide resistance in plants by expression of a detoxifying enzyme. EMBO (Eur Mol Biol Organ) J 6:2513–2518

Delannay X, LaVallee BJ, Proksch RK et al 1989 Field performance of transgenic tomato plants expressing *Bacillus thuringiensis* var *Kurstaki* insect control protein. Bio/Technology 7:1265–1269

Deom CM, Oliver MJ, Beachy RN 1987 The 30-kilodalton gene product of tobacco mosaic virus potentiates virus movement. Science (Wash DC) 237:389–394

Dingwall C, Dilworth SM, Black SJ, Kearsey SE, Cox LS, Laskey RA 1987 Nucleoplasmin cDNA sequence reveals polyglutamic acid tracts and a cluster of sequences homologous to putative nuclear localization signals. EMBO (Eur Mol Biol Organ) J 6:69–74

Domoney C, Casey R 1984 Storage protein precursor polypeptides in cotyledons of *Pisum sativum* L. Identification of, and isolation of a cDNA clone for, an 80,000-M$_r$ Legumin-related polypeptide. Eur J Biochem 139:321–327

Domoney C, Casey R 1987 Changes in legumin messenger RNAs throughout seed development in *Pisum sativum* L. Planta (Berl) 170:562–566

Domoney C, Ellis THN, Davies DR 1986 Organization and mapping of legumin genes in *Pisum*. Mol & Gen Genet 202:280–285

Flavell RB, Goldsbrough AP, Robert LS, Schnick D, Thompson RD 1989 Genetic variation in wheat HMW glutenin subunits and the molecular basis of breadmaking quality. Bio/Technology 7:1281–1285

Forster A, Symons RH 1987 Self-cleavage of virusoid RNA is performed by the proposed 55-nucleotide active site. Cell 50:9–16

Forster A, Davies C, Sheldon C, Jeffries A, Symons RH 1988 Self-cleaving viroid and newt RNAs may only be active as dimers. Nature (Lond) 334:265–267

Fraley RT, Rogers SG, Horsch RB et al 1983 Expression of bacterial genes in plant cells. Proc Natl Acad Sci USA 80:4803–4807

Garcia-Olmedo F, Salcedo G, Sanchez-Monge R, Gomez L, Royo J, Carbonero P 1987 Plant proteinaceous inhibitors of proteinases and α-amylases. Oxf Surv Plant Mol Cell Biol 4:275–334

Gatehouse AMR, Gatehouse JA, Dobie P, Kilminster AM, Boulter D 1979 Biochemical basis of insect resistance in *Vigna unguiculata*. J Sci Food Agric 30:948–958

Gerlach WL, Llewellyn D, Haseloff J 1987 Construction of a plant disease resistance gene from the satellite RNA of tobacco ringspot virus. Nature (Lond) 328:802–805

Grist DH 1986 Rice. Longman, New York, p 381

Halford NG, Forde J, Anderson OD, Greene FG, Shewry PR 1987 The nucleotide and deduced amino acid sequences of an HMW glutenin gene from chromosome 1B of bread wheat *Triticum aestivum* L and comparison with those of genes from chromosomes 1A and 1D. Theor Appl Genet 75:117–126

Hartings H, Maddaloni M, Lazzaroni N, Di Fonzo MM, Salamini F, Thompson R 1989 The *O2* gene which regulates zein deposition in maize endosperm encodes a protein with structural homologies to transcriptional activators. EMBO (Eur Mol Biol Organ) J 8:2795–2801

Haseloff J, Gerlach WL 1988 Simple RNA enzymes with new and highly specific endoribonuclease activities. Nature (Lond) 334:585–591

Haughn GW, Smith J, Mazur B, Somerville C 1988 Transformation with a mutant *Arabidopsis* acetolactate synthase gene renders tobacco resistant to sulfonylurea herbicides. Mol & Gen Genet 211:266–271

Hemenway C, Fang R-X, Kaniewski WK, Chua NH, Tumer NE 1988 Analysis of the mechanism of protection in transgenic plants expressing the potato virus X coat protein or its antisense RNA. EMBO (Eur Mol Biol Organ) J 7:1273–1280

Hiatt A, Cafferkey R, Bowdish K 1989 Production of antibodies in transgenic plants. Nature (Lond) 342:76–78

Hilder VA, Barker RF, Samour RA, Gatehouse AMR, Gatehouse JA, Boulter D 1989 Protein and cDNA sequences of Bowman-Birk protease inhibitors from the cowpea (*Vigna unguiculata* Walp.). Plant Mol Biol 13:701–710

Hoekema A, Huisman MJ, Molendijk L, van den Elzen PJM, Cornelissen BJC 1989 The genetic engineering of two commercial potato cultivars for resistance to potato virus X. Bio/Technology 7:273–278

Hoffman LM, Ma Y, Barker RF 1982 Molecular cloning of *Phaseolus vulgaris* lectin mRNA and use of cDNA as a probe to estimate lectin transcript levels in various tissues. Nucleic Acids Res 10:7819–7828

Hoffman LM, Donaldson DD, Herman EM 1988 A modified storage protein is synthesized, processed, and degraded in the seeds of transgenic plants. Plant Mol Biol 11:717–729

Karlin-Neumann GA, Tobin E 1986 Transit peptides of nuclear-encoded chloroplast proteins share a common amino acid framework. EMBO (Eur Mol Biol Organ) J 5:9–13

Katagiri F, Lam E, Chua N-H 1989 Two tobacco DNA-binding proteins with homology to the nuclear factor CREB. Nature (Lond) 340:727–730

Klein TM, Gradziel T, Fromm ME, Sanford JC 1988 Factors influencing delivery into *Zea mays* cells by high-velocity microprojectiles. Bio/Technology 6:559–563

Lawrence MC, Suzuki E, Yarghese JN et al 1990 The three-dimensional structure of the seed storage protein phaseolin at 3Å resolution. EMBO (Eur Mol Biol Organ) J 9:9–15

Lawton MA, Tierney MA, Nakamura I et al 1987 Expression of a soybean β-conglycinin gene under the control of the cauliflower mosaic virus 35S and 19S promoters in transformed petunia tissues. Plant Mol Biol 9:315–324

Lindow SE, Arny DC, Upper CD 1978 Distribution of ice nucleation-active bacteria on plants in nature. Appl Environ Microbiol 36:831–838

Ling KC 1972 Rice virus diseases. Intl Rice Res Inst, Los Baños, Phillipines, p 95

Loesch-Fries LS, Merlo D, Zinnen T et al 1987 Expression of alfalfa mosaic virus RNA4 in transgenic plants confers virus resistance. EMBO (Eur Mol Biol Organ) J 6:1845–1851

Lyon BR, Llewellyn DJ, Huppatz JL, Dennis ES, Peacock WJ 1989 Expression of a bacterial gene in transgenic tobacco plants confers resistance to the herbicide 2,4-dichlorophenoxyacetic acid. Plant Mol Biol 13:533–540

Matta NK, Gatehouse JA, Boulter D 1981 Molecular and subunit heterogeneity of legumin of *Pisum sativum* L. (garden pea)—a multidimensional gel electrophoretic study. J Exp Bot 32:1295–1307

Moreno J, Chrispeels MJ 1989 A lectin gene encodes the α-amylase inhibitor of the common bean. Proc Natl Acad Sci USA 86:7885–7889

Neuhaus G, Spangenberg G, Mittelsten Scheid O, Schweiger H-G 1987 Transgenic rapeseed plants obtained by the microinjection of DNA into microspore-derived embryoids. Theor Appl Genet 7:30–36

Neurath H 1989 Proteolytic processing and physiological regulation. Trends Biochem Sci 14:268–271

Newmark P 1987 Trypsin inhibitor confers pest resistance. Bio/Technology 5:426

Osborn TC, Alexander DC, Sun SM, Cardona C, Bliss FA 1988 Arcelin is a lectin-homologous seed protein that confers insect resistance. Science (Wash DC) 240: 207–210

Palmenberg A, Pallansch MA, Rueckert RR 1979 Protease required for processing picornaviral coat protein resides in the viral replicase gene. J Virol 32:770–778

Paulson JC 1989 Glycoproteins: what are the sugar chains for? Trends Biochem Sci 14:272–276

Payne PI 1987 Genetics of wheat storage proteins and the effect of allelic variation on bread-making quality. Annu Rev Plant Physiol 38:141–153

Potrykus I 1988 Direct gene transfer to plants. In: Applications of plant cell and tissue culture. Wiley, Chichester (Ciba Found Symp 137) p 144–162

Potrykus I 1990 Gene transfer methods for plants and cell cultures. In: Bioactive compounds from plants. Wiley, Chichester (Ciba Found Symp 154) p 198–212

Powell PA, Stark DM, Sanders PR, Beachy RN 1989 Protection against tobacco mosaic virus in transgenic plants that express tobacco mosaic virus antisense RNA. Proc Natl Acad Sci USA 86:6949–6952

Powell-Abel P, Nelson RS, De B et al 1986 Delay of disease development in transgenic plants that express the tobacco mosaic virus coat protein gene. Science (Wash DC) 232:738–743

Rezaian MA, Skene KGM, Ellis JG 1988 Antisense RNAs of cucumber mosaic virus in transgenic plants assessed for control of the virus. Plant Mol Biol 11:463–471

Romero J 1984 Genetic variability in the seed protein of non-domesticated bean (*Phaseolus vulgaris* L var aborigeneus) and the inheritance and physiological effects of arcelin, a novel seed protein. PhD thesis, University of Wisconsin, Madison

Shah DM, Horsch RB, Klee HJ et al 1986 Engineering herbicide tolerance in plants. Science (Wash DC) 233:478–481

Sheehy RE, Kramer M, Hiatt WR 1988 Reduction of polygalacturonase activity in tomato fruit by antisense RNA. Proc Natl Acad Sci USA 85:8805–8809

Smith CJS, Watson CF, Ray J et al 1988 Antisense RNA inhibition of polygalacturonase gene expression in transgenic tomatoes. Nature (Lond) 334:724–726

Stalker DM, McBride KE, Malyj LD 1988 Herbicide-resistance in transgenic plants expressing a bacterial detoxification gene. Science (Wash DC) 242:419–423

Stark DM, Beachy RN 1989 Protection against potyvirus infection in transgenic plants: evidence for broad spectrum resistance. Bio/Technology 7:1257–1262

Stussi-Garaud C, Garaud JC, Berna A, Godefroy-Colbuen T 1987 In situ localization of an alfalfa mosaic virus non-structural protein in plant cell walls: correlation with virus transport. J Gen Virol 68:1779–1784

Tabata T, Takase H, Takayama S et al 1989 A protein that binds to a cis-acting element of wheat histone genes has a leucine zipper motif. Science (Wash DC) 245:965–967

Tatham AS, Miflin BJ, Shewry PR 1985 The β-turn conformation in wheat gluten proteins: relationship to gluten elasticity. Cereal Chem 62:405–412

Tatham AS, Drake AF, Field JM, Shewry PR 1987 The conformation of three synthetic peptides corresponding to repeat motifs of two cereal prolamins. In: Lasztity R, Bekes F (eds) Proceedings 3rd International Workshop on glutenin proteins. World Scientific, Singapore, p 460–496

Tumer NE, O'Connell KM, Nelson RN et al 1987 Expression of alfalfa mosaic virus coat protein gene confers cross-protection in transgenic tobacco and tomato plants. EMBO (Eur Mol Biol Organ) J 6:1181–1188

Vandekerckhove J, Van Damme J, Van Lijsebettens Botterman J et al 1989 Enkephalins produced in transgenic plants using modified 2S seed storage proteins. Bio/Technology 7:929–932 van der Krol AR, Lenting PE, Veenstra J et al 1988a An anti-sense chalcone synthase gene in transgenic plants inhibits flower pigmentation. Nature (Lond) 333:866–869

van der Krol A, Mol J, Stuitje A 1988b Antisense genes in plants: an overview. Gene 72:45–50

Van Dun CMP, Bol JF, van Vloten-Doting L 1987 Expression of alfalfa mosaic virus and tobacco rattle virus coat protein genes in transgenic tobacco plants. Virology 159:299–305

Van Dun CMP, Bol JF 1988 Transgenic tobacco plants accumulating tobacco rattle virus coat protein resist infection with tobacco rattle virus and pea early browning virus. Virology 167:649–652

Voelker TA, Herman EM, Chrispeels MJ 1989 In vitro mutated phytohemagglutinin genes expressed in tobacco seeds: role of glycans in protein targeting and stability. Plant Cell 1:95–104

Wilcox DR, Shivakuma AG, Melin BE et al 1986 Genetic engineering of bioinsecticides. In: Inouye M, Sarma R (eds) Protein engineering. Academic Press, New York, p 395–383

Wolf S, Deom CM, Beachy RN, Lucas WJ 1989 Movement protein of tobacco mosaic virus modifies plasmodesmatal size exclusion limit. Science (Wash DC) 246:377–379

DISCUSSION

MacMillan: You mentioned a transgenic rice plant—what variety of rice was this?

Hall: Many groups are working with rice. The model we use is Taipei 309, a japonica type. With support from Texas state funds and industry, we are going to apply these techniques to the indica varieties. Several of the Texas lines are elite germplasm. The Chulabhorn Research Institute here in Bangkok is also interested in transgenic rice, as are many third world countries.

The key aspect is that to produce transgenic rice plants that are fertile appears to add a dimension of difficulty. Shimamoto et al (1989) in Japan have second generation plants.

Bellus: In the field of insect-resistant plants, the major work was done with *Bacillus thuringiensis*. There are some good transgenic tobacco plants which have resistance against insects. You mentioned very briefly that there are still some problems with other plants because the mRNA is not stable or there are other problems with transformation of plants.

Hall: Transformation and regeneration are a major barrier for rice but this looks like being broken very soon. Many lepidopteran insects cause severe losses of crops.

van Montagu: B. thuringiensis has not only been used to make tobacco resistant to insects. Potatoes have been made resistant to potato tuber moth by overexpressing the BT gene. Several other plants (e.g. tomatoes, lettuce) have been engineered for such resistance. Other possibly more important crops, such as cotton, are being worked on.

Barz: Dr Hall, the work on the herbicide bromoxynil is wonderful and we all agree that this strategy opens immense possibilities. However, I would like to look at some possible side effects of the strategy. You said that when the gene for the bacterial nitrilase is put into a plant, the herbicide is degraded. This is not quite true; you just hydrolyse the nitrile group. The product generated by this reaction is a *p*-hydroxybenzoic acid substituted with two bromine atoms. This is an ideal substrate for peroxidative polymerization into lignin and other 'bound residues'. To have bromine-substituted aromatic compounds in these bound residues raises ecological or possible toxicological considerations. The original herbicide is not incorporated into the bound residues, whereas the product is.

I would suggest that you should not incorporate a nitrilase, but rather express or overexpress a suitable glutathionetransferase to remove the bromine atoms from the molecule. That would give the same result, inactivation of the herbicide, but you would not generate a substrate which can be incorporated into the bound residues. Would you agree with me?

Hall: Your point is a valuable one and deserves serious consideration. I don't have all the details of the toxicology; this is a commercial project and Rhône-Poulenc have gone through all the toxicological tests. In addition, this herbicide is applied at an extremely low rate.

Rickards: The nitrilase experiments and the hydrolysis of the herbicide nitrile to the carboxylic acid might have a secondary benefit. The acid produced is already known as a natural product that has significant antifungal activity.

van Montagu: If you can, it is better to detoxify the herbicide. Monsanto and Calgene have been expressing an altered target enzyme for the herbicide, 'Roundup'. They have found that the engineered plants looked healthy after having been sprayed with the herbicide, but seed production was decreased. This was because some cell types, in this case in the reproductive tissues, didn't express the altered target enzyme at a sufficiently high level.

A section through the anthers of a transgenic plant that was expressing the glucuronidase gene under the control of a specific promoter showed that the gene was expressed only in one cell type, e.g. the tapetum or the stomium or the endothecium.

Hall: This is an important point, that regulation is not just in the tissues, it occurs in every single cell. It is easy when you look at a tissue, for example the mesophyll cells in rice, to say that the β-glucuronidase gene is expressed throughout the tissue, but in fact each cell is under its own regulation. There may be transport from one cell to another of various molecules, but each cell is individually regulated. This is important in the herbicides, because if there is one particular cell type in a leaf that is not protected against the herbicide, the herbicide may kill those cells, resulting in easy access for pathogens.

Kuć: Dr Beachy's group has constitutively expressed the coat protein of tobacco mosaic virus, cucumber mosaic virus and other viruses in tobacco, tomato and other plants; this is extremely important. It has caused an explosion in thought and in publications. But this is a relatively simple system and there's still a great deal of controversy as to the mode of action—how the constitutive expression of the coat protein protects the plant. For a while it was thought that it prevented uncoating of the introduced virus but there are some objections to this explanation.

Could you suggest how we should proceed with more complex systems, namely diseases caused by bacteria or fungi?

Hall: The coat protein protection is simplistic and remarkable in its breadth. I believe the coat protein simply blocks the receptors for the virus. Although these have yet to be identified in plants, it seems that there have to be receptors on the leaf for these viruses, so basically one saturates receptors by coat protein. This approach does not protect against viral nucleic acid. If you take the naked RNA from these viruses and rub it on a leaf, you produce an infection. If the leaf has a wound, you will still get an infection.

With the satellite virus approach, I showed that one can interfere with the replication system by introducing a small region of RNA which we believe binds to the viral replicase very tightly, thereby preventing viral replication. So with regard to your question, if we can find a part of the life cycle of a fungus or a bacterium that requires something encoded by the invading organism, and use, for example, enzyme kinetics to get to a binding constant that does not allow release, this approach may be widely applicable. We are at the beginning of these techniques for protection against bacteria and fungi.

MacMillan: A subject of general concern to those interested in plants and plant products is plant growth and how this is regulated. I just want to say a couple of things about the regulation of plant growth by gibberellins. I will talk only about maize but we have done similar things with the dicot, pea.

There are single gene dwarf varieties of maize that are gibberellin deficient. They cannot make gibberellins and they grow normally only if they are treated with gibberellins. Using these mutants, B. O. Phinney and I have determined the biosynthetic pathway for gibberellins in maize shoots and have shown the positions of all of the blocks in the biosynthetic pathway in these mutants. We have bioassayed all the precursors and the gibberellins and shown that there is only one bioactive gibberellin, GA_1. This controls the stem elongation. There is a minor pathway which goes to gibberellin A_3.

We have a crucial enzyme which hydroxylates the inactive precursor, giberrellin A_{20}, to an active gibberellin, GA_1, for plant growth. The second pathway, to gibberellin A_3, seems to be controlled by the same enzyme. We have isolated an enzyme from another system, seeds of *Phaseolus vulgaris*, which has both these activities on the one enzyme—hydroxylation and dehydrogenation. This enzyme system is very interesting to those who want to study growth regulation of plants.

Other enzymes have been isolated that have 2β-hydroxylase activity. They deactivate gibberellins and are therefore important in the regulation of plant growth.

We have raised a monoclonal antibody to GA_1. From that monoclonal antibody we have made an anti-idiotypic monoclonal antibody which recognizes the receptor. We have now shown where that receptor is in one plant system. So this is where we stand on the hormonal regulation of plant growth.

Hall: These pathways are very exciting. This relates to the questions about regenerability of crops. One of the great needs is to understand the genes regulating pathways, such as the gibberellic acid pathway and cytokinin pathways. That insight may be the key to understanding regulation. If different tissues express different members of gene families, there may be a developmentally regulated cascade which is working in the opposite direction to that of applied growth hormones. So the sort of systems that you described appear to be very valuable.

MacMillan: They are tissue specific. I described the stem situation. In the seed of pea the situation is quite different.

Reference

Shimamoto K, Terada R, Izawa T, Fujimoto H 1989 Fertile transgenic rice plants regenerated from transformed protoplasts. Nature (Lond) 338:274–276

Gene transfer methods for plants and cell cultures

Ingo Potrykus

Institute for Plant Sciences, Swiss Federal Institute of Technology (ETH), ETH-Zentrum, LFW-E59.1 CH-8092 Zurich, Switzerland

Abstract. Agrobacterium-mediated gene transfer provides a routine and efficient gene transfer system for a variety of plant species. As this biological vector does not, however, function with important plant species, numerous alternative approaches have been studied. Of those, direct gene transfer into protoplasts, microinjection and biolistics have been demonstrated to be effective. Others, for example, viral vectors, agroinfection, liposome injection and electrophoresis may have special merits, although transgenic plants have not been produced by these techniques yet. From methods based on pollen transformation, the pollen tube pathway, pollen maturation, incubation of dry seeds, incubation of tissues, liposome fusion with tissues, macroinjection, laser treatment and electroporation of tissues no proof of integrative transformation is available, so far, and it is difficult to envisage how these approaches will ever produce transgenic cells and plants. We discuss (a) why *Agrobacterium* does not function with all plants, (b) what merits and disadvantages we see for the effective methods, (c) what possibilities we foresee for some of the other approaches, and (d) why we do not expect the remaining ones to be successful.

1990 Bioactive compounds from plants. Wiley, Chichester (Ciba Foundation Symposium 154) p 198–212

Gene transfer to plants is normally achieved via the biological vector *Agrobacterium tumefaciens*. As this system does not function with important species and varieties, alternative approaches have been studied. Some of those have been successful (e.g. biolistics and microinjection) and one has been developed to a routine and efficient method (direct gene transfer). Others are still in the exploratory stage and some may not have much potential at all. We have established direct gene transfer for cereals and focus now on microinjection into pro-embryos with the aim of developing a gene transfer protocol that can be applied to every plant species. In the following paper, however, we shall present an assessment of all approaches tested so far.

Agrobacterium tumefaciens

This soil bacterium has the capacity to transfer DNA which is flanked by two border sequences to plant cells, where the DNA integrates into the host genome.

This biological phenomenon has been elucidated in detail and exploited for the development of efficient, simple and routine gene transfer protocols. *Agrobacterium*-mediated DNA transfer works *in vivo* and *in vitro*, with plants, seedlings, tissue explants, cell cultures and protoplasts; it has led to the regeneration of numerous transgenic plants from a variety of species (Rogers & Klee 1987, Zambryski et al 1989). *Agrobacterium* would be a perfect vector for gene transfer to plants, if it was effective in all plant species. Unfortunately, this is not the case. Many important crop plants, especially the cereals, do not respond to this natural vector. What is the problem between *Agrobacterium* and cereals? It was believed until recently that there was no interaction between cereal cells and *Agrobacterium* and that consequently there was no DNA transfer. However, elegant experiments have shown that *Agrobacterium* probably is able to transfer DNA into cereal cells (Grimsley et al 1987). Why then have no transgenic cereals been recovered from the numerous treatments of cereal tissues with engineered *Agrobacterium*? I am afraid that the lack of cooperation lies on the cereal side and that it has a biological reason which cannot be altered: *Agrobacterium* requires for its function a specific state of competence in the plant cell. This state is normally built up in wound-adjacent cells as a reaction to the 'wound response' in dedifferentiating differentiated cells (Potrykus 1990a,b). Differentiated cereal cells apparently lack the wound response and (consequently?) the capacity for dedifferentiation. This is probably also the case for other important plant species. Even if DNA could be transferred into wound-adjacent cells in these plants and even if it integrated, this would not lead to transgenic cell clones because the wound-adjacent cells die and do not proliferate. The few proliferation-competent cells in meristems are either not accessible to infection or not very accessible to transformation. As there is probably little hope that *Agrobacterium* can be used for genetic engineering of several important plants, alternative methods have to be explored.

Direct gene transfer into protoplasts

Plants can be regenerated from protoplasts—isolated, single plant cells that have lost their cell walls (Potrykus & Shillito 1986). Protoplasts can be made from a great variety of plant tissues, including *in vitro* cell cultures and organ cultures. As protoplasts are surrounded by only a simple plasma membrane, it was possible to introduce genes directly with straightforward physical treatments (Paszkowski et al 1984). Within a few years the basic method was optimized such that transgenic plants could be produced with great efficiency either by electroporation (Shillito et al 1985) or by simple chemical treatments (Negrutiu et al 1987). These plants showed perfect Mendelian inheritance of the foreign genes (Potrykus et al 1985) with gene stabilities comparable to those of original plant genes. This method of 'direct gene transfer' also enabled efficient co-transformation of plants (Schocher et al 1986) and gene targeting (Paszkowski et al 1988). Cereal protoplasts could be transformed early in these studies but it took a few years longer for transgenic cereals to be regenerated (Shimamoto et al 1989, Rhodes et al 1989). It is probably correct to predict that there is no limitation concerning gene transfer to protoplasts and that plants which can be regenerated from protoplasts can therefore be transformed. There are about 150 species of which plants have been regenerated (Roest & Gilissen 1989) and additional species where cell cultures have been recovered from protoplasts. Is this the solution of the problem then? Unfortunately, it is probably not. Plant regeneration from protoplasts is a very delicate and often unreliable process depending upon too many parameters that are not under experimental control.

Particle gun or biolistics

This method has been developed from the rather surprising idea of bypassing all biological complications by shooting genes glued to heavy particles into plant cells (Klein et al 1987). Explosive forces of gunpowder or heated water are used to accelerate large numbers of small metal particles into target tissues. This has led to transient expression of marker genes. The first case of regeneration of transgenic plants as a consequence of biolistic treatment of shoot meristems came from the important plant species soybean (McCabe et al 1988). Since then numerous laboratories using biolistic devices in large-scale experiments have found that this technology is not as efficient as had been expected. It is even rather inefficient as far as integrative transformation is concerned. What about its function with plant species that cannot be transformed by other methods? It has already been stated that the only transgenic cereals come, so far, from protoplasts and direct gene transfer. It is surprising that it is taking so long for the first proven case of a transgenic cereal to be recovered from biolistic treatment. Why is the biolistic approach so inefficient and why has it not yet yielded transgenic cereals? I see several possible causes and it will be difficult to alter them experimentally: (a) the frequency of integrative transformation depends upon the concentration of the transforming gene; if this falls below a critical level, the transformation frequency is close to zero. It may be difficult to deliver enough DNA on a particle or the DNA may be released too slowly from the particle. (b) Recovery of a transgenic plant requires that the foreign gene is transported to and integrates into a cell which is competent for integrative transformation and for clonal propagation and regeneration. Such cells are probably extremely rare in those plants which cannot be transformed by *Agrobacterium*. (c) Even if successful, integrative transformation in one cell of a multicellular structure will yield a transgenic chimaera and if no secondary morphogenesis from, and selection for, those transgenic cells is easily possible, it will depend on good fortune whether or not a transgenic sector will contribute to the 'germline'. As, however, so much effort has been invested in the biolistic approach, we can expect that sooner or later novel transgenic plants will become available. I am, however, not at all convinced that biolistics can be developed into an efficient routine technique for every plant species. Biolistics would be the ideal gene transfer method for cell cultures because millions of units can be targeted with one shot (and *Agrobacterium* is rather inefficient in cell cultures), if it yielded higher frequencies of integrative transformation.

Microinjection

There is another technique which looked very promising at the beginning but which since then has encountered surprising problems. It was demonstrated three years ago that microinjection of a marker gene into microspore-derived pro-embryos produced transgenic chimaeras (Neuhaus et al 1987). This technique uses microscopic devices and microinjection capillaries to deliver defined volumes of DNA solutions into individual cells without impairing their viability. We have made every effort to repeat this success with cereals and to extend it to zygotic pro-embryos. We believed that the key problems would be related to the production of enough microspore-derived pro-embryos in our laboratory from the main cereals, and to the isolation of very young zygotic pro-embryos and regeneration of plants from them. Both problems were solved: many hundreds of microspore-derived pro-embryos of maize, wheat, rice and barley and of zygotic pro-embryos from these cereals have been microinjected and regenerated, and many thousands of sexual offspring have been analysed for the presence of the foreign gene. Surprisingly, we have, so far, no case where we can prove that the foreign gene has been inherited by the offspring. What are the problems with this approach? One problem is certainly the same as with the biolistic approach, the chimaeric nature of putative transgenic plants which makes transmission to offspring a chance event. We should, however, also consider the possibility that cereal pro-embryos and meristems do not contain many cells competent for integrative transformation. The key to the development of an efficient gene transfer method for cereals and other plants may not be the method of DNA delivery, but the availability of competent cells. This is where protoplasts, probably, have one of their great advantages over all other systems: protoplast isolation shifts potentially competent cells into the competent state.

As none of the four approaches discussed so far has led to the development of a generally and routinely applicable gene transfer method, and as all of them have their inherent problems, it is still worth discussing briefly the other approaches which have not yielded proven integrative transformation, even with easier plant systems. Before discussing the various approaches, it is necessary to define what constitutes 'proof of integrative transformation'. Many years of experience with artifacts in numerous laboratories have shown that genetic, phenotypic or physical data alone are not acceptable. Proof requires (a) serious controls for treatments and their subsequent analysis, (b) tight correlation between treatment and predicted results, (c) tight correlation between physical and phenotypic data, (d) *complete* Southern analysis containing the predicted signals in high molecular weight DNA, hybrid fragments between host DNA and foreign gene, the complete foreign gene and evidence for the absence of contaminating DNA fragments; (e) data which allow discrimination between false positives and true positives in the evaluation of the phenotypic evidence; (f) if possible, correlation of the phenotypic and physical evidence with transmission to sexual offspring, as well as genetic and molecular analysis of offspring populations. Judged on this basis, none of the following approaches has, so far, provided proof of integrative transformation.

DNA and RNA viral vectors

Replacement of an unessential viral gene by a selectable marker gene in cauliflower mosaic virus produced, after viral infection, turnip plants which contained and expressed the marker gene throughout the plant (Brisson et al 1984). It was hoped that this principle could be extended to many genes and viruses to exploit the presence of multiple copies of target genes. Viral genomes are, apparently, so tight that they do not easily tolerate foreign DNA. Since normal cloning technology can also be applied to RNA viruses, this far larger group of plant viruses is open to a series of interesting engineering approaches (Ahlquist & French 1988). However, viral vectors, will probably not contribute much to the production of transgenic plants, owing to the obvious lack of integration into the host genome, unless somebody finds conditions for the efficient integration of viral genomes.

'Agroinfection'

The DNA transfer mechanism of *A. tumefaciens* can also be used for the transfer of viral genomes into plants ('agroinfection') where mechanical virus infection is not possible (Grimsley et al 1986). The virus then spreads systemically throughout the plant. This amplification of a single DNA transfer event has been used to study whether or not *Agrobacterium* can deliver DNA into cereals. Systemic spread of a maize DNA virus in maize after agroinfection showed that this is possible (Grimsley et al 1987). It was later shown that there is not much difference in the efficiency in DNA transfer between dicots and monocots. Was this then the long-expected proof that *Agrobacterium* is a viable vector system for genetic engineering of cereals? This is, unfortunately, most probably not the case. If one used an engineered virus carrying a foreign gene, one could spread this gene in the individual plant which was agroinfected. As the virus is excluded from the meristems, however, it is excluded from the transmission to the offspring. As the virus does not integrate into the host genome, there is little chance that it integrates into the genome of a cell of the 'germline' or one of the rare cells competent for transformation and regeneration. The only chance for integration of the foreign gene would be in wound-adjacent cells which are not competent and do not survive.

Liposome injection

Microinjection into differentiated cells can easily deposit the DNA into the vacuole, where it is degraded. Lucas et al (1990) had the idea of using the vacuole to deliver DNA to the cytoplasm. Microinjection of liposomes into the vacuole leads to fusion with the tonoplast, thereby releasing the contents of the liposome into the cytoplasm, as demonstrated with cytoplasm-activated fluorescent dyes. Activity of injected DNA has still to be shown. This method, though elegant, has probably no advantage over straightforward microinjection, especially for the production of transgenic cereals. Cereals regenerate only from meristematic cells which do not have large vacuoles.

Electrophoresis

Radioactive-labelled marker gene solutions have been electrophoresed across the shoot meristem area of barley seeds (Ahokas 1989). The author has shown by autoradiography that label moves along the cell walls. The possibility has not been excluded that the cell wall labelling is due to breakdown products. If DNA could travel such long distances in cell walls, it should also be possible to transport it across cell walls. This would be an important finding. So far, there is no proof for integrative transformation by this method.

Pollen transformation

In the early 1970s experiments were undertaken to test whether or not pollen could be used as a vehicle for the delivery of foreign genes into sexual offspring (Hess 1987). Since then numerous laboratories have performed experiments with increasing sophistication and application of molecular techniques. Although several interesting phenotypes have been recovered in the offspring, there is no proven case of gene transfer. There are, however, many clearly negative results from large-scale experiments in experienced laboratories. If it were possible to transform a zygote by incubation of germinating pollen in DNA solutions, this would be an ideal method. As this has, however, in more than fifteen years of experimentation never produced a single proven transgenic offspring, the chances that this will be developed into an efficient method are rather small.

Pollen tube pathway

It was tempting to test whether pollen tubes could be used for gene delivery to the zygote. Applications of marker gene solutions to cut pistils a few hours after pollination in rice (*Oryza sativa*) led to kanamycin-resistant offspring (not a reliable marker); Southern blots of their DNA unfortunately do not prove that integrative transformation has occurred (Luo & Wu 1988). Again, if this worked, it would represent an ideal method. However, it is difficult to reproduce the data and there are several biological problems, e.g. callose plugs within the pollen tubes, nucleases, cell wall adsorption, synergids, DNA transport, etc., so it would be surprising if this approach did work.

Pollen maturation

Three key problems may have prevented pollen transformation, so far: the cell wall, nucleases and the intense heterochromatic state of the sperm cells. The last two problems may be overcome by the approach of *in vitro* maturation (Alwen et al 1990). Immature microspores can be matured to functional pollen *in vitro*. If genes could be transferred at the microspore stage, they might have a chance to integrate.

Incubation of dry seeds

The most recent experiments use dry cereal embryos separated from the kernel at the scutellum, thereby creating a large wound. Incubation in viral or non-viral DNA solutions yielded evidence for transient expression of marker genes and recombination of viral DNA (Töpfer et al 1989). It is hoped that regeneration of plants from these embryos will lead to transgenic cereals and a large number of offspring is under study. There is, so far, no proof of integrative transformation and the chances for transgenic cereals are extremely small: use of viral DNA may lead to systemic spread which will not lead to integration; non-viral DNA has virtually no opportunity to reach competent cells that will contribute to shoot regeneration. The best this DNA can do is to reach some of the wound-adjacent cells, but these do not proliferate, and they die.

Incubation of tissues or cells

There have been many approaches where seedlings, organs, tissues, cells or cell cultures of numerous plant species have been brought into direct contact with foreign DNA and defined marker genes. Treatment also included experimental designs making use of open plasmodesmata or loosening of cell wall structures. There were also treatments ensuring that competent cells were available at a sufficiently high frequency. Even in experiments which would have recovered extremely rare events of integrative transformation, there was not a single proven case of such transformation. Experiments relying on the passage of functional genes across cell walls have very little chance of success, not only because the cell wall is a perfect barrier against large DNA molecules, but also because it is an efficient trap. Even if there were occasional transfer, there are other negative parameters which act in a multiplicative way: (a) attachment to cell walls, (b) transport across further cell walls, (c) no mechanism for DNA transport, (d) competent cells have to be reached. The combination of several low frequency events will cause problems, even if one step may work occasionally (Potrykus 1990a,b).

Liposome fusion with tissues and protoplasts

Fusion of DNA-containing liposomes with protoplasts is an established method for the production of transgenic plants (Deshayes et al 1985). It has, however, no obvious advantage over direct gene transfer. DNA-containing liposomes have also been applied to various tissues, cell cultures and pollen tubes, with the rationale that liposomes might help transport the DNA via plasmodesmata or directly across the cell wall. Liposomes can carry small dye molecules into cells within tissues via fusion with the plasmalemma (Gad et al 1988). There is, however, no proof of transport and integration of marker genes. As plasmodesmata are sealed off immediately after wounding, this route is not open even for small liposomes; impregnation of the cell wall with phospholipids seems not to change its barrier function.

Macroinjection

Use of injection needles with diameters greater than cell diameters leads to destruction of those cells into which DNA is delivered. DNA integration would require that the DNA moves into cells adjacent to wounds and, therefore, all the problems discussed before apply. The most exciting data, so far, were from an experiment where a marker gene was injected into the stem below the floral meristem of rye (*Secale cerealse*) (De la Pena et al 1987). Hybridization to the marker gene and enzyme assays with selected sexual offspring yielded strong indicative evidence. Unfortunately, it has so far not been possible either to reproduce these data in several large-scale experiments with other cereals or to establish proof with the original material. It is very difficult to understand how the DNA would reach the sporogenic cells in this experimental design, as DNA would have not only to reach neighbouring cells but also to travel across many layers of cells.

Microlaser

A microlaser beam focused into the light path of a microscope can be used to burn holes in cell walls and membranes (Weber et al 1988). It was hoped that incubation of perforated cells in DNA solutions could serve as a basis for a vector-independent gene transfer method into walled plant cells. There are no conclusive data available on DNA uptake and there are problems with adsorption of exogenous DNA to cell wall material, even before it can be taken up. As microinjection and biolistics definitely transfer DNA into walled plant cells, the microlaser would offer advantages only in very specific cases where those techniques were not applicable.

Electroporation into tissues

Discharge of a capacitor across a cell containing protoplasts in a DNA solution is a routine method for DNA uptake and transient, as well as stable, integrative transformation (Shillito et al 1985, Fromm et al 1986). Electroporation has also been applied to a variety of walled plant cells and tissues. The results confirm that cell walls are very efficient barriers against DNA molecules of the size of a functional gene.

Summary

Gene transfer to plants is routine and efficient with *A. tumefaciens* for plant species with a prominent wound response and with 'direct gene transfer' for species which can be regenerated from protoplasts. Transgenic cereals (rice, maize) have been recovered, so far, exclusively from the latter method. This method is not applicable to all plant species. The challenge, therefore, is to develop a routine gene transfer method which achieves integrative transformation with cereals and other important crop plants, or with any given plant species and variety. Two approaches have the best potential in this respect: biolistics and microinjection into zygotic pro-embryos. It cannot be excluded, of course, that the other approaches may still cause some surprises. For cell culture transformation, direct gene transfer may be the method of choice if there are no problems with protoplast isolation and culture. Biolistics is relatively easy to apply and efficient in the production of transient signals; however, it is still very inefficient for stable integrative transformation. This is, unfortunately, also true for the use of *Agrobacterium* with many cell cultures, although efficient transformation of cell cultures has been reported (Pollock et al 1985). Microinjection, of course, is possible, but requires the most sophisticated instrumentation.

References

Ahlquist P, French R 1988 Molecular genetic approaches to replication and gene expression in brome mosaic and other RNA viruses. In: Domingo E et al (eds) RNA genetics, vol III. Variability of RNA genomes. CRC Press, Boca Raton, p 53–69
Ahokas H 1989 Transfection of germinating barley seed electrophoretically with exogenous DNA. Theor Appl Genet 77:469–472
Alwen A, Eller N, Kastler M, Benito-Moreno RM, Heberle-Bors E 1990 Potential of in-vitro pollen maturation for gene transfer. Physiol Plant 79:194–196

Brisson N, Paszkowski J, Penswick JR, Gronenborn B, Potrykus I, Hohn T 1984 Expression of a bacterial gene in plants by using a viral vector. Nature (Lond) 310:511–514

De la Pena A, Lörz H, Schell J 1987 Transgenic plants obtained by injecting DNA into young floral tillers. Nature (Lond) 325:274–276

Deshayes A, Herrera-Estrella L, Caboche M 1985 Liposome-mediated transformation of tobacco mexophyll protoplasts. EMBO (Eur Mol Biol Organ) J 4:2731–2737

Fromm ME, Taylor LP, Walbot V 1986 Stable transformation of maize after electroporation. Nature (Lond) 319:791–793

Gad AE, Zeewi B, Altmann A 1988 Fusion of germinating watermelon pollen tubes with liposomes. Plant Sci 55:68–75

Grimsley NH, Hohn G, Hohn T, Walden R 1986 Agroinfection, an alternative route for viral infection of plants using the Ti plasmid. Proc Natl Acad Sci USA 83: 3282–3286

Grimsley NH, Hohn T, Davies JW, Hohn B 1987 Agrobacterium-mediated delivery of infectious maize streak virus into maize plants. Nature (Lond) 325:177–179

Hess D 1987 Pollen based techniques in genetic manipulation. Int Rev Cytol 107: 169–190

Klein TM, Wolf ED, Wu R, Sanford JC 1987 High-velocity microprojectiles for delivering nucleic acids into living cells. Nature (Lond) 327:70–73

Lucas WJ, Lansing A, de Wet JR, Walbot V 1990 Introduction of foreign DNA into walled plant cells via liposomes injected into the vacuole: a preliminary study. Physiol Plant 79:184–189

Luo Z, Wu R 1988 A simple method for transformation of rice via the pollen-tube pathway. Plant Mol Biol Rep 6:165–174

McCabe DE, Swain WF, Martinell BJ, Christou P 1988 Stable transformation of soybean (Glycine max) by particle acceleration. Bio/Technology 6:923–926

Negrutiu I, Shillito RD, Potrykus I, Biasini G, Sala F 1987 Hybrid genes in the analysis of transformation conditions. I. Setting up a simple method for direct gene transfer in plant protoplasts. Plant Mol Biol 8:363–373

Neuhaus G, Spangenberg G, Mittelsten Scheid O, Schweiger HG 1987 Transgenic rapeseed plants obtained by microinjection of DNA into microspore-derived proembryoids. Theor Appl Genet 75:30–36

Paszkowski J, Shillito RD, Saul M et al 1984 Direct gene transfer to plants. EMBO (Eur Mol Biol Organ) J 3:2717–2722

Paszkowski J, Baur M, Bogucki A, Potrykus I 1988 Gene targeting in plants. EMBO (Eur Mol Biol Organ) J 7:4021–4027

Pollock K, Barfield DG, Robinson SJ, Shields R 1985 Transformation of protoplast-derived cell colonies and suspension cultures by Agrobacterium tumefaciens. Plant Cell Rep 4:202–205

Potrykus I 1990 Gene transfer to cereals: an assessment. Bio/Technology 8:535–542

Potrykus I 1990 Gene transfer to plants: assessment and perspectives. Physiol Plant 79:125–134

Potrykus I, Shillito RD 1986 Protoplasts: isolation, culture, plant regeneration. In: Weissbach A, Weissbach H (eds) Methods in Enzymology, vol 118: Plant molecular biology. Academic Press, Orlando, p 549–578

Potrykus I, Paszkowski J, Saul MW, Petruska J, Shillito RD 1985 Molecular and general genetics of a hybrid foreign gene introduced into tobacco by direct gene transfer. Mol Gen Genet 199:169–177

Rhodes CA, Pierce DA, Mettler IJ, Mascarenhas D, Detmer JJ 1989 Genetically transformed maize plants from protoplasts. Science (Wash DC) 240:204–207

Roest S, Gilissen LJW 1989 Plant regeneration from protoplasts: a literature review. Acta Bot Neerl 38:1–23

Rogers SG, Klee H 1987 Pathways to plant genetic manipulation employing *Agrobacterium*. In: Hohn T, Schell J (eds) Plant gene research, plant DNA infectious agents. Springer-Verlag, New York, p 179–203

Schocher RJ, Shillito RD, Saul MW, Paszkowski J, Potrykus I 1986 Co-transformation of unlinked foreign genes into plants by direct gene transfer. Bio/Technology 4:1093–1096

Shillito RD, Saul MW, Paszkowski J, Müller M, Potrykus I 1985 High frequency direct gene transfer to plants. Bio/Technology 3:1099–1103

Shimamoto K, Terada R, Izawa T, Fujimoto H 1989 Fertile rice plants regenerated from transformed protoplasts. Nature (Lond) 338:274–276

Töpfer R, Gronenborn B, Schell J, Steinbiss HH 1989 Uptake and transient expression of chimeric genes in seed-derived embryos. Plant Cell 1:133–139

Weber G, Monajembashi S, Greulich KO, Wolfrum J 1988 Injection of DNA into plant cells with a UV laser microbeam. Naturwissenschaften 75:35–36

Zambryski P, Tempé J, Schell J 1989 Transfer and function of T-DNA genes from *Agrobacterium* Ti and Ri plasmids in plants. Cell 56:193–201

DISCUSSION

Hall: May I ask, Ingo, what line of indica rice you were using?

Potrykus: We used Chinsurah Boro 2 (Datta et al 1990).

van Montagu: We have used electroporation to introduce DNA into intact leaf tissue of rice (Dekeyser et al 1990). When we introduced a glucuronidase gene under the control of the cauliflower mosaic virus 35S promoter, we could detect expression of this gene in most of the cell types in the rice leaf tissue. These experiments indicated that the transfer of plasmid DNA to the electroporated cells is rather uniform. When we electroporated a glucuronidase construct under control of the *S*-adenosylmethionine synthetase promoter, we could detect glucuronidase activity only in the vascular tissues of rice. This expression pattern is very similar to that of the same construct in transgenic *Arabidopsis* or tobacco plants. We have also obtained expression of neomycin phosphotransferase activity using this method. Therefore, we think that electroporation of intact tissues will help studies on promoters and tissue-specific gene expression in rice. The method also works in wheat and in barley.

Potrykus: I am very excited about this. Many people have tried seriously to electroporate DNA into tissues and failed. We spent one year trying to find evidence for transformation after electroporation of various experimental systems. Nobody up to this moment has shown that DNA crosses cell walls. If the interpretation of your data is correct, and (so far) everything speaks for that, this would be a major advance for gene technology. It would then be no problem to electroporate DNA into meristems, from which plants could be regenerated, and this would facilitate the recovery of transgenic plants from many species for which this has not yet been achieved.

My problem with these data is that I know of many experiments where this approach was completely negative. I do not trust staining for glucuronidase because there have been many false positives with this method. The data which Marc described, however, do not rely only on glucuronidase staining; there is additional evidence, e.g. the neomycin phosphotransferase activity. The data also show organ-specific expression, as expected from the promoter used. However, I still worry that there is an artifact leading to a misinterpretation.

van Montagu: I have to stress that so far we have only introduced DNA into explants of graminaceous plants by this method. Initial experiments with tobacco have not yet yielded positive results. We have also tried to regenerate stably transformed tissue, so far without success.

Hall: Marc, are those very young leaves?

van Montagu: This is mainly done on tissue from seven day old seedlings.

MacMillan: Concerning the transformation of plants, Dr Potrykus drew our attention to the fact that one of the difficulties is the loss of competence for regeneration. Is it possible to manipulate this? Has it been done, for example by the use of indole acetic acid or other plant cytokinins?

Potrykus: I myself have tested over a period of six years more than 120 000 different variations of experimental treatments, including a large series of plant growth regulators. I have tested 164 different auxin derivatives across a wide range of concentrations and in up to seven factor combinations. So far there is no way to correct this apparent loss of totipotency, despite great experimental efforts.

Cragg: This work is very impressive but what are the prospects for enhancing the production of complex secondary metabolites, for example, a drug like taxol, which is showing tremendous promise as an antitumour agent?

Potrykus: We have the technology at the moment for enhanced production of every gene product for which the gene has been isolated. The question is how much is known about the genetic basis of secondary metabolites. What are the chances of isolating the genes involved? This is the key problem—how long will it take to identify the underlying biochemistry, the key proteins, and use those to isolate the essential genes?

Battersby: Is there a single case known where a reasonably complex secondary metabolite has been increased substantially in quantity by genetic manipulation?

Potrykus: I think this is coming only now. The delay is probably in part due to the fact that our colleagues interested in secondary metabolites only recently found the connection to molecular biology!

van Montagu: People working in secondary metabolites have always tried to purify these enzymes and that can take many years. The techniques are now available to clone the genes from cell fractions that are simply enriched in these enzymes. Even if there are still 500 proteins in your mixture, if you can identify the key enzymes on a two-dimensional gel, the microsequencing techniques are sufficient to obtain protein sequence data. Then you can go for the gene. We have seen in collaboration with Professor Barz that this is possible.

The next step is to take the promoter of the genomic sequence and overproduce the enzyme. All that is known from bacteria and from yeast suggests that this will be possible with plants. What is striking for me is that we have never found any interest in this from a pharmaceutical company. We have many times asked them; the answer is always 'do it and we will see'.

Hall: This approach of identifying the enzymes and then cloning the genes is the reverse of that used with herbicides, which is to take plant genes and put them into bacterial, fungal or baculovirus expression systems, from which you can get very large amounts of the required protein. There is no reason why we can't take plant metabolites and put them into bacterial expression systems— which may be much better than the plant systems for a whole range of reasons.

The other consideration is that the targeting of these materials to places within the plant where they can be accumulated in high concentrations and remain stable may also be very important.

Fowler: Going back to yesterday's presentation by Professor Barz, the excitement I had was to see the coordinate induction of phytoalexin systems. There, to me, lies a key to Gordon Cragg's question. We are beginning to see ways in which to induce not just a single gene or a single enzyme of a pathway but a whole pathway. Then if we can isolate promoters of the sort that Marc van Montagu mentions, we can enhance the overall expression of those genes for greater product synthesis.

Yamada: We have been working on scopolamine biosynthesis (Yamada & Hashimoto 1988, Yamada et al 1990a). Many plants produce *l*-hyoscyamine and scopolamine. We purified hyoscyamine 6β-hydroxylase (Hashimoto & Yamada 1987, Yamada et al 1990b), which introduces the hydroxyl group. We found that this 2-oxoglutarate-dependent dioxygenase (hydroxylase) also catalyses the epoxidation which is the next step in the pathway from *l*-hyoscyamine to scopolamine (Hashimoto et al 1989). This enzyme is thus a multifunctional enzyme. For this purified hydroxylase, we now have a cDNA and a monoclonal antibody. We have transformed tobacco with this cDNA and it is expressed.

Dr Liu Yong-Long mentioned that 6β-hydroxyhyoscyamine is as important as *l*-hyoscyamine (p 175). This hydroxylase converts 6β-hydroxyhyoscyamine to scopolamine. It is impossible to transform plants with the genes for each enzyme in a synthetic pathway, but the enzyme that catalyses a key step can be overexpressed in a plant to increase the yield from the overall pathway.

van Montagu: The problem is that you need enough substrate for that key enzyme. The indications from Professor Barz's data were that if you know the trigger, for example the induction of phytoalexins by natural infection, you see that the whole pathway is coordinately expressed. Molecular biologists have indications that promoter regions have binding sites for many proteins which act as transcription factors. Some of these proteins are characteristic for a given pathway. If we could clone those proteins or select for mutations in these proteins, we could control the whole pathway. That is one of the projects we are pursuing at the moment. One enzyme will teach us a lot and the steps taken by Professor Yamada are extremely important. It is only step by step that we will learn. Theory is always nice but practice is the only answer.

Yamada: In this case quite a few plants produce, as the end product, *l*-hyoscyamine, which is the precursor of scopolamine, so we don't have to give any precursor to the transformed plants; we can just overexpress the hyoscyamine 6β-hydroxylase to produce 6β-hydroxyhyoscyamine and scopolamine.

Fowler: A word of caution about control systems. We may focus on an individual pathway and enhance enzyme activity, but we can't look upon a single pathway in isolation. Each pathway is in balance with the other pathways in a cell. The cell is a dynamic system, an integrated system; if you alter one pathway, the knock-on effects may be quite dramatic.

Bowers: In all these systems there are certainly feedback mechanisms that limit the production of secondary plant metabolites. There's enormous variability amongst individual plants in the amount of secondary metabolites that they produce. Could one use molecular biological techniques to diminish the effectiveness of a rate-limiting feedback mechanism?

Hall: If the rate-limiting step is feedback onto the promoter region, by putting in a different promoter you may completely block that feedback. Clearly, we will learn a lot more about intermediary metabolism and the interactions between pathways in the near future.

Potrykus: I think there is no general answer. Nothing is as convincing as a successful experiment and we have just heard of two successful experiments. I assume that within the next year a number of interesting experiments will be done, because the technology is available and now it will be used.

Battersby: I agree completely. There is intense interest from bioorganic chemists and from those working on secondary plant metabolites in approaches based on gene transfer technology.

Bellus: I would be very interested to hear some predictions. Let's say we have a specific problem: a compound of natural origin which is very active biologically and has 11 asymmetric centres. We need 100 kg to do all the necessary toxicity studies. So where should we invest our money? In a conventional plantation or in molecular biology or in bioorganic chemistry? And we would like to have the 100 kg within one year!

Potrykus: If you need it in one year, I certainly would invest in the plantation.

Balick: As your financial adviser, I would go for the plantation.

Farnsworth: I always give this answer—if the cocaine cartel in Colombia can produce unlimited quantities of the secondary metabolite that seems so fascinating, then it can't be such a big problem with any other plant.

Barz: I think for the next five years, Dr Bellus, you should invest in plantations, but not to 100%; let's say to 95% only. You should give the other

5% to molecular biology, and within about 5–10 years you will give 100% to molecular biology.

Ley: I think you will find that organic chemists will be using molecular biology as well. I don't see any divisions between molecular biologists, organic chemists or chemists of any type. We are all scientists and we shall be using all these methods simultaneously.

References

Datta SK, Peterhans A, Datta K, Potrykus I 1990 Genetically engineered fertile Indica-rice recovered from protoplasts. Bio/Technology 8:736–740

Dekeyser RA, Claes B, De Rycke RMU, Habets ME, van Montagu MC, Caplan AB 1990 Transient gene expression in intact and organized rice tissues. Plant Cell 2:591–602

Hashimoto T, Yamada Y 1987 Purification and characterization of hyoscyamine 6β-hydroxylase from root cultures of *Hyoscyamus niger* L. Eur J Biochem 164:277–285

Hashimoto T, Kohno J, Yamada Y 1989 6β-Hydroxyhyoscyamine epoxidase from cultured roots of *Hyoscyamus niger*. Phytochemistry (Oxf) 28:1077–1082

Yamada Y, Hashimoto T 1988 Biosynthesis of tropane alkaloids. In: Applications of plant cell and tissue culture. Wiley, Chichester (Ciba Foundation Symposium 137) p 199–212

Yamada Y, Hashimoto T, Endo T et al 1990a Biochemistry of alkaloid production *in vitro*. In: Charlwood BV (ed) Secondary products from plant tissue culture. Oxford University Press, p 225–240

Yamada Y, Okabe S, Hashimoto T 1990b Homogeneous hyoscyamine 6β-hydroxylase from cultured roots of *Hyoscyamus niger*. Proc JPN Acad Ser B Phys Biol Sci 66:73–76

Compounds from plants that regulate or participate in disease resistance

Joseph Kuć

Department of Plant Pathology, University of Kentucky, Lexington, Kentucky 40546, USA

Abstract. Disease resistance is multifactorial. The response phase includes: synthesis of phytoalexins, i.e. low molecular weight antimicrobial compounds which accumulate at sites of infection; systemically produced enzymes which degrade pathogens, e.g. chitinases, β-1,3-glucanases and proteases; systemically produced enzymes which generate antimicrobial compounds and protective biopolymers, e.g. peroxidases and phenoloxidases; biopolymers which restrict the spread of pathogens, e.g. hydroxyproline-rich glycoproteins, lignin, callose; and compounds which regulate the induction and/or activity of the defence compounds, e.g. elicitors of plant and microbial origin, immunity signals from immunized plants and compounds which release immunity signals.

Disease resistance in plants is not determined by the presence or absence of genes for resistance mechanisms, it is determined by the speed and degree of gene expression and the activity of the gene products. It is likely, therefore, that all plants have the genetic potential for resistance. This potential can be expressed systemically (immunization) after restricted inoculation with pathogens, attenuated pathogens or selected non-pathogens, or treatment with chemical substances that are produced by immunized plants or chemicals which release such signals. Immunization is effective against diseases caused by fungi, bacteria and viruses, and it has been successfully tested in the laboratory and field.

Advances in science have provided information and technology to enhance resistance to plant pests. Pesticides are part of this technology, but they also contribute to a complex world problem which threatens our environment and hence our survival. The future will see the restriction of pesticide use and a greater reliance on resistant plants generated using immunization and other biological control technologies, genetic engineering and classical plant breeding. However, as with past and current technology, we will have created unique problems. The survival of our planet depends upon anticipating these problems and meeting the challenge of their solution.

1990 Bioactive compounds from plants. Wiley, Chichester (Ciba Foundation Symposium 154) p 213–228

Plants are protected against disease caused by infectious agents by multiple mechanisms that act at different stages of infection (Kuć 1985, 1987, 1990, Carr & Klessig 1989, Dean & Kuć 1987). External plant surfaces are often covered by biopolymers that make penetration difficult. These polymers include waxes, and cutin and suberin, which are fatty acid esters. In addition, external tissues can be rich in phenolic compounds, alkaloids, diterpenoids, steroid glycoalkaloids and other compounds which inhibit the development of fungi and bacteria (Kuć 1985, Carr & Klessig 1989, Reuveni et al 1987). Cell walls of at least some monocotyledons also contain antimicrobial proteins, referred to as thionins (Carr & Klessig 1989, Bohlman et al 1988). It is unclear whether the amount of these proteins increases as a result of infection.

Once they have penetrated the external barriers or entered through wounds, pathogens may encounter plant cells containing sequestered glycosides. When such cells are ruptured by injury or infection, they release the glycosides, which may be antimicrobial *per se* or may be hydrolysed by glycosidases to yield antimicrobial phenols; these in turn may be oxidized to highly reactive, antimicrobial quinones and free radicals (Kuć 1985, Noveroske et al 1964, Dean & Kuć 1987). Thus, damage to a few cells may rapidly create an extremely hostile environment for a developing pathogen. This rapid, but restricted, disruption of a few cells after infection can also result in the biosynthesis and accumulation of low molecular weight antimicrobial, lipophilic compounds, called phytoalexins, in and around the site of infection. They differ widely in structure, with some structural similarities within plant families (Kuć 1985, Carr & Klessig 1989, Bailey & Mansfield 1982, Dean & Kuć 1987). Some are synthesized by the malonate pathway, others by the mevalonate or shikimate pathways, whereas still others require participation of two or all three of the pathways (Fig. 1). Since phytoalexins are not translocated, their protective effect is limited to the region of the infection, and their synthesis and its regulation are accordingly restricted. Phytoalexins are degraded by some pathogens and by the plant; thus they are transient constituents and their accumulation is a reflection of both synthesis and degradation.

Often associated with phytoalexin accumulation is the deposition around sites of injury or infection of biopolymers, which both mechanically and chemically restrict further development of pathogens (Kuć 1985, Carr & Klessig 1989, Dean & Kuć 1987, Hammerschmidt & Kuć 1982, Rao & Kuć 1990). These biopolymers include: lignin, a polymer of oxidized phenolic compounds; callose, a polymer of β-1,3-linked glucopyranose; hydroxyproline-rich glycoproteins, and suberin.

The macromolecules produced after infection or certain forms of physiological stress include enzymes which can hydrolyse the walls of some pathogens (Carr & Klessig 1989, Boller 1987, Rao & Kuć 1990), including chitinases, β-1,3-glucanases and proteases. Unlike the phytoalexins and structural biopolymers, the amounts of these enzymes increase systemically in infected plants even in response to localized infection. They are often found intercellularly where they would contact fungi and bacteria. These enzymes are part of a group of stress or infection-related proteins commonly referred to as pathogenesis-related (PR) proteins. The function of many of these proteins is unknown. Some may

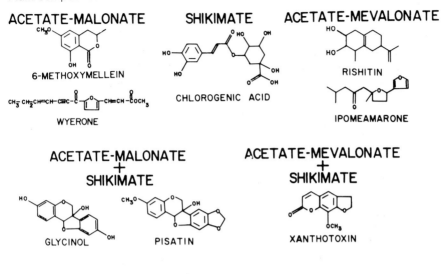

FIG. 1. Pathways utilized for the biosynthesis of some phytoalexins.

be defence compounds; others may regulate the response to infection (Carr & Klessig 1989, Boller 1987, Ye et al 1989, Tuzun et al 1989, Rao & Kuć 1990).

Another group of systemically produced biopolymer defence compounds comprises the peroxidases and phenoloxidases (Hammerschmidt et al 1982, Rao & Kuć 1990). Both can oxidize phenols to generate protective barriers to infection, including lignin. Phenolic oxidation products can also cross-link to carbohydrates and proteins in the cell walls of plants and fungi to restrict further microbial development (Stermer & Hammerschmidt 1987). Peroxidases also generate hydrogen peroxide, which is strongly antimicrobial. Associated with peroxidative reactions after infection is the transient localized accumulation of hydroxyl radicals and superoxide anion, both of which are highly reactive and toxic to cells.

Clearly disease resistance is multifactorial, as would be expected for plants to survive the selection pressure of evolution (Fig. 2). The expression of genes that encode products which contribute to disease resistance is regulated by both plant and microbial compounds. It is this regulation of gene expression, rather than the presence or absence of genes for resistance mechanisms, that determines disease resistance in plants. All of the defence compounds I have discussed can be produced by resistant and susceptible plants, even plants reported to lack genes for resistance. The speed, magnitude and timing of different elements of the response, and the activity of the gene products as influenced by a plant's internal and external environment, determine resistance/susceptibility. The ability to immunize plants systemically, including those reported to lack genes for resistance, supports the contention that resistance is not determined by the presence of the genes that code for the defence compound (Dean & Kuć 1987, Rao & Kuć 1990, Kuć 1985, 1987, 1990). Immunization has been accomplished in our laboratory by restricted inoculation with pathogens, attenuated pathogens, selected non-pathogens, and treatment with chemical substances which are signals produced by immunized plants or chemicals which release such signals (Kuć 1987, Doubrava et al 1988, Gottstein & Kuć 1989). Immunization is systemic, and effective against fungi, bacteria and viruses in the laboratory and field. Consequences of immunization include enhanced systemic activities of chitinases, β-1,3-glucanases and peroxidases and appearance of PR proteins. The plant is also sensitized to respond rapidly after infection, e.g. by phytoalexin

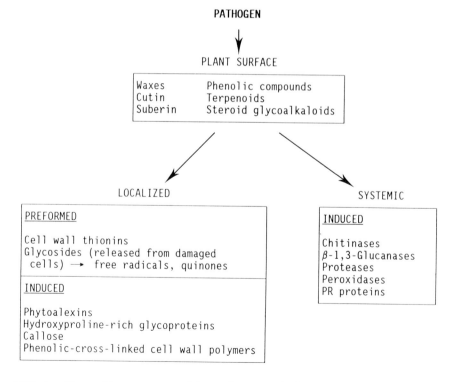

FIG. 2.   The multicomponent and layered nature of disease resistance in plants.

accumulation, lignification, further increased activity of chitinases, β-1,3-glucanases, peroxidases and increased levels of PR proteins (Kuć 1985, 1990, Dean & Kuć 1987, Ye et al 1989, Tuzun et al 1989). Cell wall polysaccharides from fungi and plants induce some of the defence responses e.g. phytoalexin accumulation and lignification, but their effect is localized and protection is not systemic (Bailey & Mansfield 1982, Rao & Kuć 1990, Kuć 1987, 1990).

Considering the varied mechanisms for resistance, it is not surprising that resistance is the general rule and susceptibility is the exception. Susceptibility however, does occur and perhaps more attention should be directed to understanding how the successful pathogen avoids or suppresses the defences of the host. If we understood the basis for susceptibility, we would gain insights useful for the development of new and effective disease control strategies.

Defence compounds for resistance

A logical consequence of the information presented concerning defence compounds would be to use them to protect plants against disease. One scenario would be to spray plants, treat seeds or soak soil with these naturally occurring compounds. The assumptions are that the compounds are safe because they occur naturally, and they would be effective because existing plants have survived the selection pressure of evolution. Though it is justified to look for such compounds in Nature, especially in parts of the world where there is great plant diversity, there are flaws in this reasoning. Naturally occurring compounds are not necessarily safe or effective. Many of the world's most potent poisons are derived from plants. Some plant tissues are extremely toxic to animals because they contain protective compounds, e.g. potato, tomato and tobacco foliage, potato peel, potato sprouts. The phytoalexins are not notably antimicrobial, though they do accumulate at sites of infection to levels sufficient to inhibit development of some fungi and bacteria. A direct role for phytoalexins in restricting virus multiplication or movement in plants is not evident. Since phytoalexins are not translocated, plants would have to be sprayed frequently to compensate for losses due to rainfall or degradation by microorganisms and plants, and to protect emerging foliage or developing fruit. Phytoalexins on seed surfaces would probably be unstable once seeds were planted in soil; soil drenches would prove unsatisfactory for the same reason. Hydrolytic enzymes would also require frequent applications. It would be unrealistic to coat plant surfaces completely with a water-insoluble polymer such as lignin. The economic considerations of using naturally occurring defence compounds to protect plants would also be unfavourable. The structure of most of these compounds is rather complex, synthesis would be difficult and the cost of isolation from natural sources would probably be high.

Having listed the flaws in the scenario, the possibility exists that natural defence compounds may be found which are highly active, can be easily and economically synthesized and are safe to use. Even if such a compound could not be used directly, it might serve as a model for the synthesis of new pesticides. If we use these compounds as pesticides, however, we reintroduce all the problems associated with the current use of pesticides. I suggest it is better to study plant defence compounds to understand the natural mechanisms for disease resistance rather than for their use in disease control. There are more promising approaches to the problem. Why not regulate the expression of plant genes and enable plants to protect themselves?

Regulation of the expression of genes for resistance

An alternative to the direct use of defence compounds to protect plants is the elicitation of their synthesis within the plant. Some cell wall components from fungi, bacteria and plants are highly active elicitors of phytoalexin accumulation (Kuć 1985, 1987, 1990, Dean & Kuć 1987, Rao & Kuć 1990). However, the use of such elicitors as foliar sprays proved ineffective, for many of the reasons that sprays containing phytoalexins prove ineffective. In addition, the repeated application of elicitors of phytoalexin accumulation caused a marked diversion of energy and carbon precursors from vital processes, resulting in resistant but stunted plants. Plants which constitutively accumulate phytoalexins were also stunted and non-productive. In Nature, phytoalexins are produced only at or around sites of infection, when needed.

Methods have been suggested to enhance resistance in plants by incorporating genes for phytoalexin synthesis from unrelated plants. The reasoning is that phytoalexins from a carrot, which is not susceptible to diseases of bean, would protect bean against its pathogens. The suggestions are unlikely to prove useful, because phytoalexins have a broad spectrum of biological activity. Those from carrot inhibit bean pathogens as well as pathogens of carrot and *vice versa*. The pathways for phytoalexin synthesis are also quite complex. Thus, it is unlikely that the transfer of a single gene would enable a carrot to produce phytoalexins found in bean.

Immunization systemically enhances the levels of some defence compounds; however, its main effect is to sensitize the plant to subsequent infection. Restricted inoculation of the first true leaf of cucumber, muskmelon and watermelon with *Colletotrichum lagenarium*, *Pseudomonas lachrymans* or tobacco necrosis virus systemically protects the roots, stems and leaves against 13 diseases caused by fungi, bacteria and viruses (Kuć 1985, 1987, 1990, Dean & Kuć 1987, Hammerschmidt & Kuć 1982, Hammerschmidt et al 1982, Rao & Kuć 1990). Protection against the three classes of pathogens is induced by inoculation with any one and persists for 4–6 weeks. A booster inoculation during this period extends protection throughout the growing season. Tobacco is systemically immunized against blue mould caused by *Peronospora tabacina* by injecting the stems of young plants with sporangiospores of the pathogen, or by inoculating 2–3 leaves of a cultivar containing the NN gene for resistance to tobacco mosaic virus (TMV) with TMV. Systemic protection against TMV is induced by inoculation with either *P. tabacina* or TMV.

Immunization in the cucurbits and tobacco is transmissible by grafting and plants regenerated via tissue culture from immunized tobacco remained immunized in laboratory and field tests (Kuć 1985, 1987, 1990, Tuzun & Kuć 1987). Three compounds isolated from immunized tobacco plants, β-ionone, 3-isobutyroyl-β-ionone and 3-*n*-butyroyl-β-ionone, protected tobacco plants against blue mould in laboratory and field tests when injected into the stem or sprayed onto foliage (Salt et al 1986, 1988). The 3-*n*-butyroyl-β-ionone is one of the most biologically active compounds reported; it inhibits the germination of *P. tabacina* spores at concentrations as low as parts per 10^{12}. Other synthetic β-ionone derivatives are active as low as parts per 10^{15}. In spite of the high inhibitory activity of the esters of 3-hydroxy-β-ionone against sporangiospore germination, their systemic protective effect is more likely to be as immunity signals which regulate the expression of resistance mechanisms rather than as inhibitors *per se*. A low molecular weight compound has also been isolated from cucumber which may function as an immunity signal (N. Garas & J. Kuć, unpulished work 1989, L. Matthews & J. Kuć, unpublished work 1989).

The immunity signals are synthesized at the site of infection and are probably transported in the phloem throughout the plant (Kuć 1985, 1987, 1990). The signals induce resistance, even in tissue which has not emerged from the bud, but protected, uninfected leaves do not make the signal molecules (Dalisay & Kuć 1989, Dean & Kuć 1986). Though the isolation and chemical characterization of immunity signals is essential to understand the molecular basis of plant immunization, and they may find use in plant protection, a second group of compounds may be even more useful. Oxalates and di and tri potassium and sodium phosphates were recently reported to release immunity signals and to protect cucumber systemically against anthracnose (Doubrava et al 1988, Gottstein & Kuć 1989). Extracts of spinach and rhubarb, which contain high levels of oxalates, also induced systemic resistance in cucumber. The oxalates and phosphates did not inhibit the growth or sporulation of *C. lagenarium*, the pathogen causing anthracnose, at concentrations considerably higher than those used to elicit systemic resistance (Doubrava et al 1988, Gottstein & Kuć 1989). Recently, the biological spectrum of effectiveness for the oxalates and phosphates in cucumber was demonstrated to include all diseases in which *C. lagenarium*, *P. lachrymans* or tobacco necrosis virus was the inducing agent (Mucharromah & J. Kuć, unpublished work). The advantage of using compounds which release immunity signals is that they convert each leaf to which they are applied into a factory leaf which generates a translocatable signal.

Immunization has provided protection against blue mould in field tests conducted in the USA, Puerto Rico and Mexico. It has also increased the yield of tobacco in the presence or absence of blue mould (Tuzun & Kuć 1989). In Mexico, immunization was very effective against strains of *P. tabacina* which were resistant to metalaxyl, the most common systemic pesticide used to control blue mould (Table 1). Spray applications of 3-*n*-butyroyl-β-ionone also provided excellent protection in field tests conducted in Mexico in 1988 (Tuzun & Kuć 1989, S. Tuzen et al, unpublished work 1989). Since immunization activates or sensitizes for activation the broad spectrum of defence mechanisms present in plants, its effects are likely to be stable, in contrast to the systemic fungicides presently in use that act at a single metabolic site to which resistant strains of pathogens develop. Unlike plant breeding for resistance, which often incorporates genes from less productive plant lines, immunization utilizes existing high-yielding, high-quality cultivars. Perhaps one direction for the future is breeding plants for ease of immunization after infection.

Plant breeding—new technologies for an old science

The first line of defence against plant disease is the development of disease-resistant plants and the utilization of natural resistance mechanisms. The plant breeder does not need to prevent extinction of all plants in the environment—adequate resistance for survival is the rule in Nature. His aim is to minimize losses from key diseases of food or fibre crops and ornamental plants from which we expect more than survival, i.e. high yield and high quality. Without disease-resistant plants, the world's population would be forced to return to the level of 'hunter–gatherer' for its food. Pesticides as we know them could not substitute for resistant plants, because their increased use would be uneconomical, increase pressure for the evolution of resistant strains of pathogens, and increase the dangers to the environment and hence survival of this planet. Though antibiotics are extremely valuable to help control bacterial diseases in animals, resistance is dependent on the immune system.

Outstanding new developments in plant breeding have utilized somaclonal variation, gametoclonal variation, interspecific or intergeneric variation, breeding in the haploid state and genetic variation from outside the plant kingdom. The insertion of the TMV coat protein gene into the tobacco genome,

TABLE 1   Injection of stems with sporangiospores of *Peronospora tabacina* immunizes tobacco against blue mould caused by the metalaxyl-sensitive and metalaxyl-resistant strains of *P. tabacina*

	Infected stems		Non-infected stems	
Treatment	Lesions per leaf	% Leaf area with lesions	Lesions per leaf	% Leaf area with lesions
Metalaxyl	0.5 ± 0.1	0.2 ± 0.1	22.1 ± 2.1	16.5 ± 1.0
Control	1.7 ± 0.3	0.1 ± 0.1	26.1 ± 2.8	17.0 ± 1.3

with the resultant development of tobacco plants resistant to TMV, has caused an explosion of information and offers promise for the development of plants resistant to diseases caused by viruses for which pesticides are not available. More recently, tobacco has been transformed with genes encoding capsid proteins from alfalfa mosaic virus, cucumber mosaic virus, potato virus X, soybean mosaic virus, and tobacco rattle virus (Carr & Klessig 1989). The number of plants transformed and viruses to which they are resistant is growing rapidly. Since the coat proteins of viruses within a group have a high degree of homology, transgenic plants expressing the coat protein gene of one virus may be resistant to disease caused by other viruses in the same group. It remains to be seen whether transgenic plants developed for resistance to viruses are useful in the field. The development of transgenic plants resistant to insects, e.g. plants producing *Bacillus thuringiensis* toxin, demonstrates the versatility of molecular biology in the development of plants resistant to multiple pests.

Classical plant breeding, however, is still vital to develop high-yielding, high-quality plants suitable for different agronomic and climatic conditions. Though the future holds promise of 'super plants' resistant to diseases, insects, herbicides, drought, heat, cold, pollution and salt injury, this promise is far from being fulfilled. In this period of great scientific and technological advancements, much of the world's population still lacks sufficient food.

Prospects for the future

One danger as we face the current technology explosion related to plant protection is that we tend to think in terms of a single technology to solve all problems, and to make the, sometimes dangerous, assumption that something new is something better. Transgenic plants have not been developed with resistance to fungal or bacterial diseases, though efforts have been and continue to be made to accomplish this. Corn plants that fix nitrogen have not been produced. Some people argue that the 'green revolution' of the recent past, which produced high-yielding crops under conditions of high soil fertility and moisture, has made the rich richer and the poor poorer.

It is important to combine technologies to take maximum advantage of each. A partially resistant plant would require less pesticide applied less frequently than would a susceptible plant, even though the resistance alone is not sufficient to protect the plant under high pathogen pressure. Immunization coupled with plants bred for resistance to some diseases would increase the level of resistance present and also the number of diseases to which the plant is resistant. Tobacco cv Tennessee 86 is resistant to etch and chlorotic vein mottling virus but highly susceptible to blue mould. Immunization of this cultivar against blue mould rapidly produces a plant that is resistant to the three diseases and the resistance can be transferred to regenerants via tissue culture (Nuckles & Kuć 1989). A transgenic plant with introduced resistance to a single disease might have a high level of resistance to some diseases bred into it and be immunized against others. Pesticide applications to such plants for disease control would be minimal.

Integrated pest management is a term well known in plant protection. The challenge is to integrate the 'old' and the 'new' to maximum advantage. This requires scientists familiar with both methodologies. It requires increased communication between the laboratory and field scientists, with the objective of clearly defining the disease problems facing agriculture and the options available to work towards their solution. A laboratory breakthrough reaches its potential when it is applied. A new conference is needed to discuss the integration of old and new technologies for plant protection. Greatly enhanced funding is needed not only for the development of new technology but also for its integration and application. For the latter, we must understand disease more thoroughly and accurately predict when it will threaten. To integrate technologies successfully requires that the product of integration will 'fit' different agronomic demands. This necessitates a new emphasis on field oriented research.

A very promising approach for disease control is the sensitization for expression of resistance mechanisms inherent in plants. Thus, plants would respond rapidly when and where needed after infection. Immunization provides such an option and it has been successfully tested in the laboratory and field. We have to reduce our dependence on pesticides for disease control and integrate all options available for plant protection. It is also important to realize that the best integrated pest management formula for the Midwest of the USA may not be the best for the Sub-Sahara region of Africa. We must be increasingly sensitive and responsive to the economic and political realities and mores of the people we serve.

As we create technology to solve problems, we create new problems. The survival of our planet depends upon our ability to anticipate the problems we create and to meet their challenge.

Acknowledgements

The author's research reported in this paper has been supported in part by grants from the National Science Foundation, Herman Frasch Foundation, CIBA-GEIGY Corporation, R. J. Reynolds Tobacco Company, and a cooperative agreement with the United States Department of Agriculture, Agricultural Research Service.

References

Bailey JA, Mansfield JW (eds) 1982 Phytoalexins. Halsted Press, New York p 334
Bohlman H, Clausen S, Behnke S et al 1988 Leaf specific thionins of barley—a novel class of cell wall proteins toxic to plant pathogenic fungi and possibly involved in the defense of plants. EMBO (Eur Mol Biol Organ) J 7:1559–1565
Boller T 1987 Hydrolytic enzymes in plant disease resistance. In: Kosuge T, Nester EW (eds) Plant-microbe interactions, molecular and genetic perspectives, vol 2. Macmillan, New York, p 385–413

Carr JP, Klessig DF 1989 The pathogenesis-related proteins of plants. In: Setlow JK (ed) Genetic engineering—principles and methods, vol 11. Plenum Press, New York & London, p 65–109

Dalisay R, Kuć J 1989 Enhanced peroxidase and its persistence in cucumber plants immunized against anthracnose by foliar inoculation with *Colletotrichum lagenarium.* Phytopathology 79:1150

Dean RA, Kuć J 1986 Induced systemic resistance in cucumber: the source of the signal. Physiol Plant Pathol 28:227–233

Dean RA, Kuć J 1987 Immunization against disease: the plant fights back. In: Pegg GF, Ayers PG (eds) Fungal infection of plants. Cambridge University Press, Cambridge, p 383–410

Doubrava NS, Dean RA, Kuć J 1988 Induction of systemic resistance to anthracnose caused by *Colletotrichum lagenarium* in cucumber by oxalate and extracts from spinach and rhubarb leaves. Physiol Mol Plant Pathol 33:69–79

Gottstein HD, Kuć J 1989 Induction of systemic resistance to anthracnose in cucumber by phosphates. Phytopathology 79:176–179

Hammerschmidt R, Kuć J 1982 Lignification as a mechanism for induced systemic resistance in cucumber. Physiol Plant Pathol 20:61–71

Hammerschmidt R, Nuckles E, Kuć J 1982 Association of peroxidase activity with induced systemic resistance in cucumber to *Colletotrichum lagenarium.* Physiol Plant Pathol 20:73–82

Kuć J 1985 Increasing crop productivity and value by increasing disease resistance through non-genetic techniques. In: Ballard R et al (eds) Forest potentials: productivity and value. Weyerhaeuser Company Press, Centralia, p 147–190

Kuć J 1987 Plant immunization and its applicability for disease control. In: Chet I (ed) Innovative approaches to plant disease control. John Wiley, New York, p 255–374

Kuć J 1990 Immunization for the control of plant disease. In: Hornby D et al (eds) Biological control of soil-borne plant pathogens. CAB International, Wallingford, p 355–373

Noveroske RL, Kuć J, Williams EB 1964 Oxidation of phloridzin and phloretin related to resistance of *Malus* to *Venturia inaequalis.* Phytopathology 54:92–97

Nuckles EM, Kuć J 1989 Induced systemic resistance to *Peronospora tabacina* in Tennessee 86 tobacco and tissue culture regenerants. Phytopathology 79:1152

Rao N, Kuć J 1990 Induced systemic resistance in plants. In: Cole GT, Hoch HC (eds) The fungal spore and disease initiation in plants and animals. Plenum Press, New York, in press

Reuveni M, Tuzun S, Cole JS, Siegel MR, Kuć J 1987 Removal of DVT from the surfaces of tobacco leaves increases their susceptibility to blue mold. Physiol Mol Plant Pathol 30:441–451

Salt S, Tuzun S, Kuć J 1986 Effects of β-ionone and abscisic acid on the growth of tobacco and resistance to blue mold. Physiol Mol Plant Pathol 28:287–297

Salt S, Reuveni M, Kuć J 1988 Inhibition of *Peronospora tabacina* (blue mold of tobacco) and related plant pathogens in vitro by esters of 3(R)-hydroxy-β-ionone. 42 Tobacco Chemists Conf. Proceedings American Tobacco Chemists Society, p 22

Stermer BA, Hammerschmidt R 1987 Association of heat shock induced resistance to disease with increased accumulation of insoluble extensin and ethylene synthesis. Physiol Mol Plant Pathol 31:453–461

Tuzun S, Kuć J 1987 Persistence of induced systemic resistance to blue mold in tobacco plants derived via tissue culture. Phytopathology 77:1032–1035

Tuzun S, Kuć J 1989 Induced systemic resistance to blue mold of tobacco. In: McKeen WE (ed) Blue mold in tobacco. Am Phytopathol Soc Press, St. Paul, p 177–200

Tuzun S, Rao N, Vogeli U, Schardl C, Kuć J 1989 Induced systemic resistance to blue mold; early induction of β-1, 3-glucanases, chitinases and other pathogenesis-related proteins in immunized tobacco. Phytopathology 79:979–983

Ye XS, Pan SQ, Kuć J 1989 Pathogenesis-related proteins and systemic resistance to blue mold and tobacco mosaic virus induced by tobacco mosaic virus, *Peronospora tabacina* and aspirin. Physiol Mol Plant Pathol 35:161–175

DISCUSSION

McDonald: You showed several ways in which you can induce this protection against pathogens. You didn't mention any sort of mechanical effects—did you try those? Do they give any protection?

Kuć: No, mechanical injury *per se* does not. That doesn't mean that if we knew how to do it correctly it would not, although we have tried many ways. We have had someone touch a leaf every three hours with dry ice or injure it mechanically.

Things such as phosphates and oxalates give a very restricted injury but they persist in the tissue. We feel that the key is a low-level but persistent stress. The organisms that we use, viruses, bacteria and fungi, cause restricted lesions and a low level of persistent stress.

Hall: These are fascinating data and very convincing. Is there any relationship to some experiments done at Rothamsted by Pierpoint et al (1977)? They also found that things like aspirin induce these pathogenesis-related proteins.

Kuć: Yes, aspirin gives us induced resistance in the leaf to which we apply it. It does not give systemic induced resistance. As Pierpoint found, it results in enhanced levels of the pathogenesis-related proteins, chitinases and β-1,3-glucanases.

Bellus: All your experiments provided prophylactic protection of the plant. Do you have some indications that you can also achieve curative action with your compounds?

Kuć: If the plant has the disease, the immunization technology—whether it's using an organism to induce resistance, or a compound that may be the signal, or a compound that releases the signal—will not have an effect. Plant immunization is preventive in nature.

Bellus: After inoculation you achieve a resistance which can last for several days and is spread to the whole plant. You explain the spreading by the presence of a signal. Can you comment on the chemical character of the signal?

Kuć: I wish I could give you the structure of the signal. In tobacco we believe there is a strong possibility that some of these β-ionone derivatives are acting as signals. When we put them on leaves, we see systemic protection. It is difficult to believe that they are acting only as antimicrobial agents *per se*. On the surface of immunized plants they could be active as antimicrobial agents preventing germination of fungal spores. However, the spores germinate and fungus does penetrate the immunized plants. Fungal growth is restricted within the plant. The β-ionone derivatives discussed do not restrict the growth of the fungus.

In the case of the cucumber we have worked on the signal for a long long time. We have had many leads, including glycoproteins. We believe now that we are dealing with a low molecular weight compound with a molecular weight of about 180. It is strongly hydrophilic. Perhaps when we have the next Ciba Foundation meeting in a related area, I will be able to tell you something about the structure. Perhaps I will have a long beard by that time as well.

Hall: You mentioned treatment with oxalic acid. I believe that in *Sclerotinia* infections oxalic acid oxidases have been invoked as being very protective. Is there any relationship to your small compound?

Kuć: We used labelled oxalic acid and it doesn't leave the leaf to which we apply it. We believe that we may be sequestering Ca^{2+} and thereby causing a limited but persistent stress. But it could be any one of many things. What's fascinating is that once we have induced resistance, we can take the inducer leaf off and, in cucumber for example, we see elevated systemic levels of peroxidase isozymes, chitinases, β-1,3-glucanases and other pathogenesis-related proteins, 42 days later. This is both in leaves that were there and in the leaves that emerged from the growing point after the inducer leaf was removed.

It's also intriguing that the induced leaf is the source of the signal, and if we induce in such a way that we grossly damage that factory leaf, we do not get protection or we get very poor protection. The key is to have enough living tissue in the inducer leaf to generate the signal, but once the signal has been generated and translocated (it appears to be translocated in the phloem) the inducer leaf is no longer necessary.

van Montagu: In a leaf if you test for the mitochondrial superoxide dismutase, you will not find it. If you add glucose, six hours later there is a high concentration of superoxide dismutase. By adding glucose you activate mitochondria and many things happen.

Fowler: I recollect that in many situations of wounding or infection, there is a major drop in the phosphate content of some of these plants. There is often an increase in bound or sequestered polyphosphate—have you observed this in these systems?

Kuć: We haven't looked for it; I am not familiar with that research.

Fowler: I was thinking in terms of a mannose system that might sequester phosphate, but the resulting molecular weight would be of the wrong order.

Potrykus: Is the signal graft transmissible in all species?

Kuć: It is graft transmissible in all plants that we have studied.

Potrykus: You mentioned that you were testing whether these acquired traits are transmitted to seeds. Do you have any results yet?

Kuć: There are two reports in the literature where induced resistance, using tobacco mosaic virus, has been reported to be seed transmissible (Roberts 1983, Wieringa-Branta & Dekker 1987). We have tried their methods and other ways to do this but we have not been successful. I'm not saying they are wrong, I am just saying it hasn't been successful in our hands.

Barz: What does it actually mean if you say that the leaves are immunized? What happens if you inoculate an immunized leaf? Do you still observe the hypersensitive response? Is it that the leaf is oversensitized so that at the slightest contact with a microorganism the hypersensitive response is activated?

Kuć: Many things happen. For example in immunized cucumber, if we are using *C. lagenarium* as the challenge, the spore will germinate, form an appressorium, but then penetrate the plant only with great difficulty. Once it has penetrated, many things happen: we observe enhanced lignification and deposition of hydroxyproline-rich glycoproteins at sites of challenge, and enhanced systemic levels of chitinases, β-1,3-glucanases and peroxidases.

In the case of green bean and the fungus *C. lindemuthianum*, let's consider two phytoalexins, kievitone and phaseollin. On the immunized plant there is a hypersensitive reaction that looks the same as the one you would see if you inoculated a bean cultivar that was resistant to a given race of the fungus with that fungus. There is the typical rapid accumulation of kievitone and phaseollin. We can take a bean cultivar that the plant breeder has said lacks genes for resistance to all races of *C. lindemuthianum* because it is susceptible to all races of the fungus. When we inoculate that bean cultivar with *C. lagenarium*, which is not a pathogen of bean, systemic resistance to all races of the bean pathogen is induced. The immunized cultivar of bean will behave exactly as a resistant variety with respect to the hypersensitive reaction and the accumulation of the two phytoalexins.

Barz: I refer to the very early stages of physical contact between the spore and the immunized leaf. We know of several cases in which spores release small necrosis-inducing proteins, and before any penetration occurs these proteins induce extremely small lesions, which are indicative of the hypersensitive response. It could be that the leaf tissue is so sensitive to such material that it only takes a few molecules to trigger this response. If you turn it around, I think the data you presented are also explicable in terms of resistance to a suppressor produced by the microorganism, which inhibits the defence reaction. If your status of immunization means an insensitivity, a lack of response to a suppressor, then you have induced resistance.

Kuć: You have captured our thinking. We've given this a great deal of thought and we feel that the immunized plant is super-sensitized to respond. Often, race specificity is lost and the plant will respond to races of pathogens to which it would not respond before immunization. There are at least three aspects to this: the enhancement of certain defence compounds, e.g. chitinases, glucanases, peroxidases and phytoalexins; sensitization of the plant such that after challenge there is a very rapid response at the site of challenge; and finally, the signal compounds that regulate all of this.

Potrykus: The question is how does a plant regulate this? It must re-set the physiology to a specific state which the plant is not normally in. There are studies of the habituation phenomenon in plants by Fred Meins where you can set plant cells into a specific state which is somatically inherited. A nice example is habituation for cytokinin independence. The interpretation was that stable feedback loops are induced. The reasoning behind these studies might help to develop a hypothesis for your case, shifting a complete organism into a specific state which is mitotically inherited and maintained over long periods.

Kuć: With tobacco, the tissue culture regenerants that are still resistant have provided us with a useful vehicle to find out which of the many things that have happened are important to induced resistance. We did not find enhanced peroxidase activity in the tissue culture regenerants. We also didn't find enhanced chitinase activity, but we did have enhanced β-1,3-glucanase activity. After challenge of the resistant regenerants with the fungus, the activity of this enzyme increases markedly. We did not see this in the plants that were regenerated from the control plants. We also found in immunized tobacco an increase in the amounts of glucose and fructose. In the regenerated plants we did not find these enhanced carbohydrate levels, therefore we reasoned that these carbohydrates really don't have a bearing on induced resistance.

Potrykus: Fred Meins also found enhanced β-1,3-glucanase activity in his habituation studies. This seems to be a general response to stress.

Barz: One brief suggestion for a possible experimental approach. It has been shown in the soybean–*Phytophthora* interaction and in the chickpea–*Ascochyta* interaction that during the early stages of elicitation a group of characteristic phosphorylated proteins is formed. They seem to be specific for the process of induction and may be involved in the transfer of the signal. I urge you to look at the pattern of phosphorylated proteins before and after elicitation.

Kuć: We are doing exactly this. We are fascinated by this process of signal transduction which somehow leads to the expression of so many unrelated genes. We get the feeling we are turning on a master switch—we are turning on the lights, the TV, the radio, the vacuum cleaner and all other things! We are also using the study of signal transduction as a backdoor way to get to the signal. If we see what is happening, we may have some further insights into what the signal could be.

Hall: Chris Cullis at the John Innes Institute had an interesting system with flax (1977). It was an environmentally induced change in the weight, height and DNA content of flax, the molecular basis of which was never resolved. There was a light condition under which flax would grow with a different morphology until it was put under a different regime. I think this was seed transmissible. There are master switch systems hidden in plants—clearly food for thought.

Bowers: Joe, have you involved your entomological colleagues to investigate whether the immunized plants have any effect on insects?

Kuć: I haven't involved them, they have become involved. Professor Dan Patten and two graduate students in the Department of Entomology are studying this. Working with melon aphids, Fall Army worms, two-spotted spider mites and greenhouse white flies, they found that immunized plants were as susceptible as controls, with no evident harmful effects on the insects. Though there was no protection against the insects, the plants were well protected against a test fungus.

McDonald: Do you have any information on the economics, particularly for the crops where the normal cultural practice is to start plants in the greenhouse and then plant out in the field?

Kuć: I believe that there is an immediate application for immunization technology with plants that are transplanted. Tobacco would be excellent for this; tomato would be another good choice. I feel the technology is generally applicable and we are exploring it in commercial greenhouses with high value crops. I feel it's applicable at various levels, even using organisms as the inducing agents. I consider immunization to be a form of biocontrol.

We also feel that the immunity signals and compounds which release the immunity signals may be used in field control of diseases. For example, phosphate is very cheap and it would be difficult to fault spraying it onto plants. In the case of the ionone derivatives, it is very difficult to get the precursor, 3-hydroxy-β-ionone, and this imposes some economic limitations. We are working on this problem. The tobacco growers in Mexico are very anxious to use the β-ionone derivatives in the field.

McDonald: I hadn't realized that the phosphate could be sprayed on.

Kuć: We can spray it on the leaves. There is always a lag time between application of inducer and observation of a systemic effect. From the time you apply the material it requires, depending on the inducer, 4–6 days. If the pathogen is in the field and there is a high inoculum density, that's not the time to immunize the plant.

References

Cullis CA 1977 Molecular aspects of the environmental induction of heritable changes in flax. Heredity 38:129–154

Pierpoint WS, Ireland RJ, Carpenter JM 1977 Modification of proteins during the oxidation of leaf phenols: reaction of potato virus X with chlorogenoquinone. Phytochemistry 16:29–34

Roberts DA 1983 Acquired resistance to tobacco mosaic virus is transmitted to the progeny of hypersensitive tobacco. Virology 124:161–163

Wieringa-Branta DH, Dekker WC 1987 Induced resistance in hypersensitive tobacco against tobacco mosaic virus by injection of intercellular fluid from tobacco plants with systemic acquired resistance. J Phytopathol (Berl) 118:165–170

Final discussion

Balick: There are something like 3–30 million species of plants and animals on the planet. In the last few days people have mentioned about 50 of them. I would like to know from the molecular biology community what really is the need for biological diversity? What percent of that 3–30 million organisms will be important at some point in the future? I ask you to be as visionary as possible; this will help us to clarify our thinking and activities in conservation.

Cragg: It has been commented several times that we need an integrated approach. Part of that integrated approach should be the conservation of biological diversity. How much is still out there which has not been investigated—how many resistant strains of food plants are there? For instance, a strain of maize has recently been discovered in Mexico that is resistant to viral diseases. How many new drugs, how many new agrochemicals are yet to be discovered?

The key to the conservation of biological diversity is the cataloguing of the species. The key to that is the training of more botanists to carry out this most important work. The National Science Foundation recently issued a report in which they have stressed this issue. They recommended that far greater resources be devoted to the training of botanists and systematists to play this key role in conservation of biological diversity.

Bowers: In support of systematics, it provides the structure in which we all work. The finest chemistry in the world would be absolutely trivial if scientists practising it could not distinguish between a pea plant and a palm tree. We all work in a framework that is essential, like mathematics. Systematics provides the framework and the evolutionary perspective. We have to know what we are working with, something about its evolution and something about the diversity that we can tap. There is no substitute for the study of systematics and the organization of plants and animals into a rational structure.

Balick: I would like an answer as either a percentage or a number, because it does affect our activities in conservation and our formation of global strategies for this work in the biotechnology area and the chemistry area—in all of the areas that we have talked about.

Potrykus: C. Tudge recently took the Devil's advocate position on this question (1989). He calculated that we could survive quite well with a total figure of different species around 50 000.

Bowers: Biological resources are something we hope to mine for ever. There is no way that anybody can put numbers on what plants or animals might provide useful information.

Cox: I was very impressed with Professor Fowler's approach yesterday in which he brought us down to earth to consider the economics of the processes and the technologies. How many drugs on the market are there today that came directly from molecular biology approaches or from transgenic plants? How does that compare to the number of drugs on the market today from traditional medicine and from studies of traditional medicinal plants?

Fowler: The answer to the first part of your question is None. It is not a fair comparison. You have to look at the organic chemistry route as well. If you go back fifty years or so there was a tremendous repertoire of drugs derived from plants. Then organic chemists became very skilful at molecular manipulation and organic synthesis took over a large part of the original natural product portfolio. Consequently, we see a shifting balance, reflecting technological development. Molecular biology is beginning to come through and we will see the balance move in this direction in the future.

Reference

Tudge C 1989 The rise and fall of Homo-sapiens-sapiens. Phil Trans R Soc Lond B 325:479–488

Summary

Alan R. Battersby

University Chemical Laboratory, Lensfield Road, Cambridge CB2 1EW, UK

This meeting has been notably interdisciplinary. The challenge of summarizing this meeting is thus a considerable one. I would like to take a number of themes that have emerged and which, to my delight, have become more and more intertwined as we have gone through this meeting.

The first theme came, very appropriately in a meeting about plants, from botanists and pharmacologists. Their message was that a wealth of knowledge of plants, plant materials and plant extracts has been accumulated by human beings over the ages and this knowledge can greatly help us now. It must be rather satisfying to our botanical colleagues to see how many of the leads in later papers originated from traditional sources. There may be doubts about the significance of quantitative data on 'hits', but those in my view are on the periphery. There was no question that this is an extremely important source of information and that information must continue to be gathered. I would emphasize that one of our jobs as researchers, as academicians and as industrialists, is to preserve, collect and generate knowledge, and this very important function was abundantly clear from the beginning.

Then we looked at synthesis. The opportunities there are enormous. It came out very strongly from a number of contributions that the decisive lead for an active compound often came from natural product structures. Such leads point to what smaller and simpler structures should be synthesized in order to obtain biologically active materials. There was a second message in this area—that highly functionalized small molecules, which may be very difficult to isolate or be produced in extremely small quantities in living systems—can, with the appropriate skills, be synthesized in quantity for pharmacological study. I believe there will be a resurgence of interest in the pharmaceutical industry and agrochemical industry in the chemistry of natural products for generating new structural leads.

The third theme was the use of the tissue cultures and cell suspensions. Here we came up against harsh economics and those were presented to us very frankly and fairly. It seems that this approach will be important for specialized usages to make small amounts of precious materials, and for production of enzymes. Also, we should not forget the importance of tissue cultures and cell suspensions for studies of biosynthesis. This was evident from a number of the shorter contributions.

The fourth theme came from many of the papers, including the one from Her Royal Highness. This was the illustration of what amazing organic chemists plants are. They can construct complex architecture carrying a wide range of functionality and rich in chirality.

A fifth theme was the approaches of molecular biology and the moving of genes into plants. Many of us had not realized that the first introduction of such a gene into a plant was as recently as 1984. Here we came up against fascinating aspects of control, promotion and expression and so on. I would like to include here the immunization aspects that we heard of from Dr Kuć. Clearly there are massive problems still to solve in that area, but it's exciting and is going to be extremely important and lively in the future. Molecular biology will provide opportunities for making not only proteins and peptides, but evidently has promise for making secondary metabolites as well.

The final theme is that of success, for this meeting has gone extremely well. There has been a lively interaction amongst the various participants and we have had discussion which was not only penetrating but brought out many critically important points.

This has been a joint meeting between the Ciba Foundation and the Chulabhorn Research Institute. This association has been hugely beneficial to the success of this meeting. I want to say, Your Royal Highness, how greatly appreciative we all are of your involvement and the involvement of the Thai scientists and your Institute in the work of this meeting, to bring it to such a high level of science.

Index of contributors

Non-participating co-authors are indicated by asterisks. Entries in bold type indicate papers; other entries refer to discussion contributions.

Indexes compiled by Liza Weinkove

Subject index